Life Cycles, Traits, and Variations

Reader

Printed in Canada

ISBN: 978-1-68380-506-9

Life Cycles, Traits, and Variations

Table of Contents

Animals Have Life Cycles

Chapter 1

Look at the two polar bears in the picture. One bear is very young. The other bear is grown. Over time, the young bear will grow and change. It will develop into an adult like the large bear.

Young animals are almost always smaller than adult animals. Over time, they grow and change into adults. Adult animals can produce young. Eventually adults age and die.

All **organisms** go through similar kinds of changes during their lives. Each stage in an animal's life, birth, growth, reproduction, and death is part of the animal's **life cycle**.

Big Question

How do different types of animals change throughout their lives?

Vocabulary

organism, n. any living thing

life cycle, n. the set of stages of an organism's life

Like all living things, polar bears grow and change throughout their lives.

Life Cycle of a Mouse

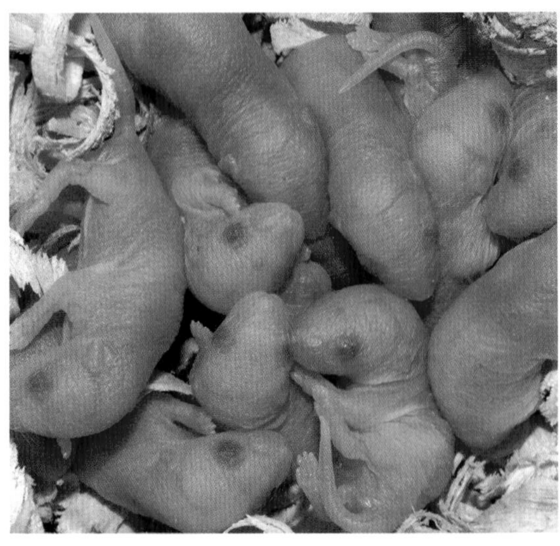

Birth: When mice are born, they are called pups. They are about the size of a quarter. They are born deaf, blind, and bald. They cannot move around by themselves. Mouse pups rely on their mother for milk. Within a week or so, they start to grow fur and teeth. In about ten days, their eyes finally open.

Vocabulary

adolescence, n. a stage of the life cycle when a young animal is developing into an adult

Growth—Adolescence: In a few weeks, the pups grow and start to change. The time when mice are developing into adults is called **adolescence**. Adolescent mice leave the nest about five weeks after birth.

Growth—Adulthood: When mice mature, they become adults. They look for mates to reproduce and start their own families.

Reproduction: Adult mice mate to make new baby mice. The formation of new young is called **reproduction**. Mice need to be fully grown adults before they can reproduce. Adult female mice do not reproduce forever though. They lose the ability to reproduce as they age.

Vocabulary

reproduction, n. the process of making new organisms

Death: When adult mice start to reach the end of their lives, they stop reproducing and eventually die.

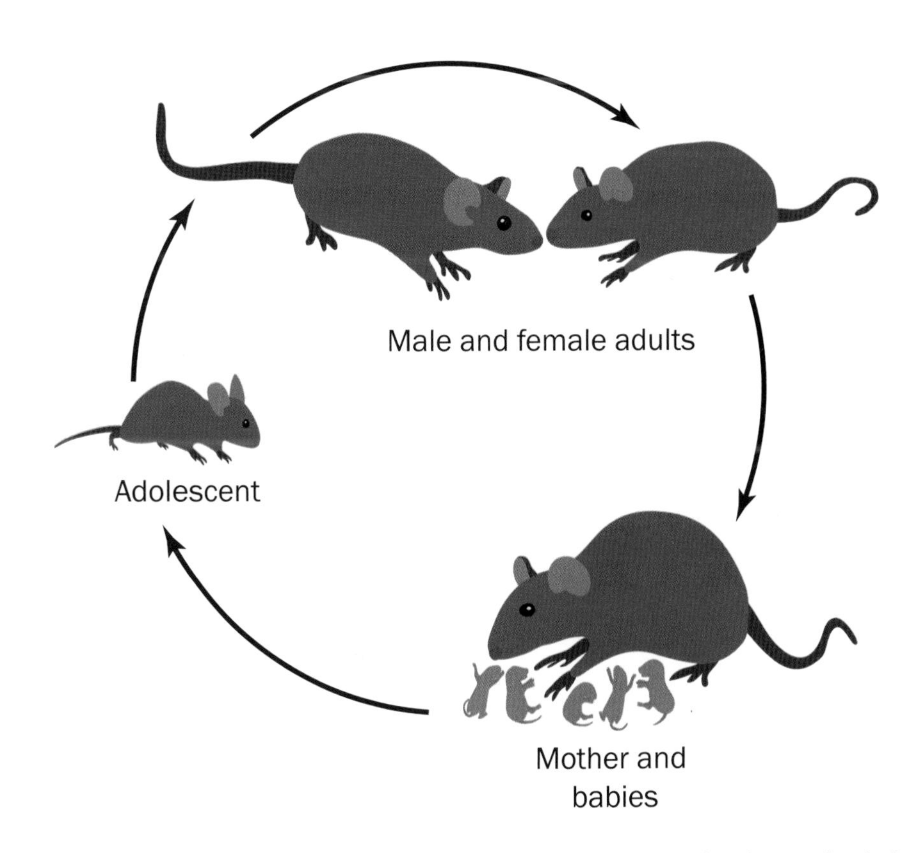

When scientists make a model of a life cycle, they often draw it in the shape of a circle to show how an animal changes during its life cycle. This model shows how a life cycle repeats for a species.

Life Cycle of a Chicken

Different types of animals reproduce in different ways. Some animals lay eggs. Eggshells protect the young animal until it is ready to live in the outside world. Chickens lay eggs as a part of their reproduction process.

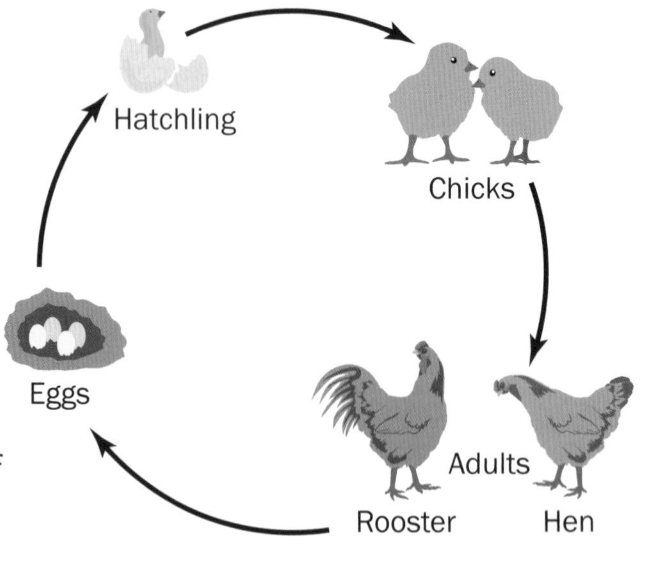

Chickens begin life in the safety of an egg. The shell protects their growth.

The chick grows until it is ready to leave the shell. It breaks out of the shell. The chick is now known as a hatchling.

Hatchlings continue to grow. They will develop into adult chickens.

Adult rooster and hen pairs can produce offspring. Hens lay eggs. Parents protect the eggs and keep them warm until they are ready to hatch.

Life Cycle of a Butterfly: Metamorphosis

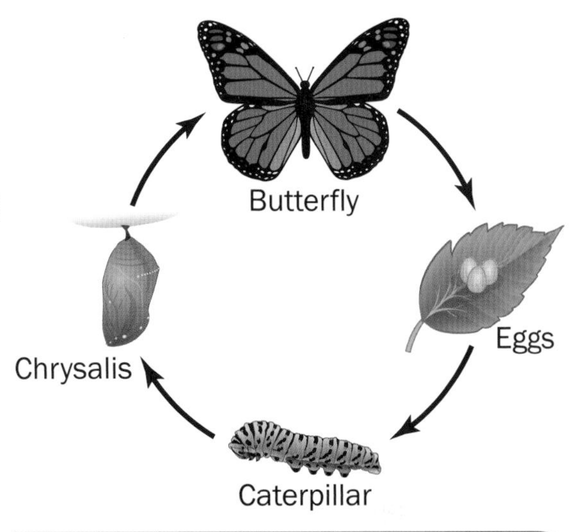

Butterflies change form during their life cycle. They develop through four stages. The stages make up a process called **metamorphosis**.

At the end of its individual life cycle, the butterfly dies. But the life cycle continues with the butterfly's young.

Vocabulary

metamorphosis, n. a change of form during the life cycle of some animals

Stage 1: Female butterflies lay eggs. The eggs are very small.

Stage 2: Caterpillars hatch from the eggs. Caterpillars are also called larvae. A larva eats and grows.

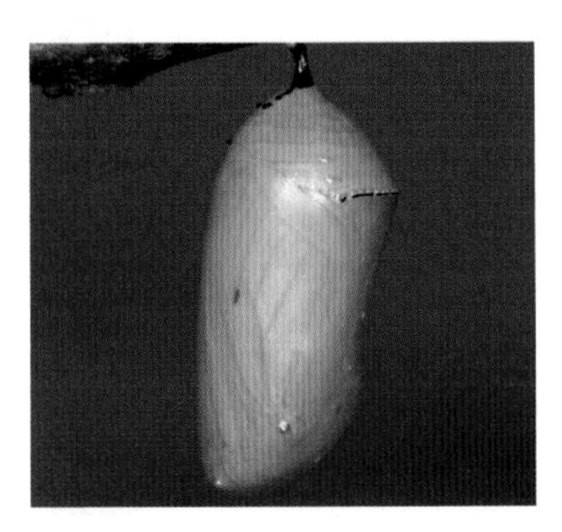

Stage 3: Each caterpillar makes a case that surrounds the caterpillar. The case is called a chrysalis or pupa. Inside, the caterpillar changes into a butterfly.

Stage 4: The butterfly comes out of the chrysalis. It is an adult. Adult butterflies mate and produce eggs.

Life Cycle of a Frog: Metamorphosis

Like butterflies, frogs change form as they grow and develop. A frog's life cycle is another example of metamorphosis.

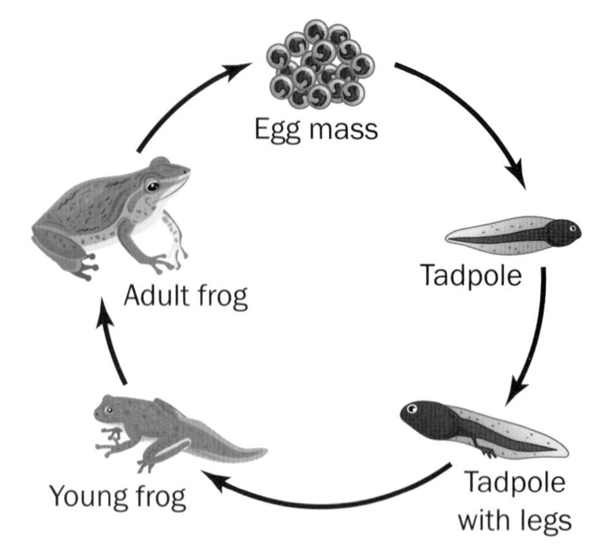

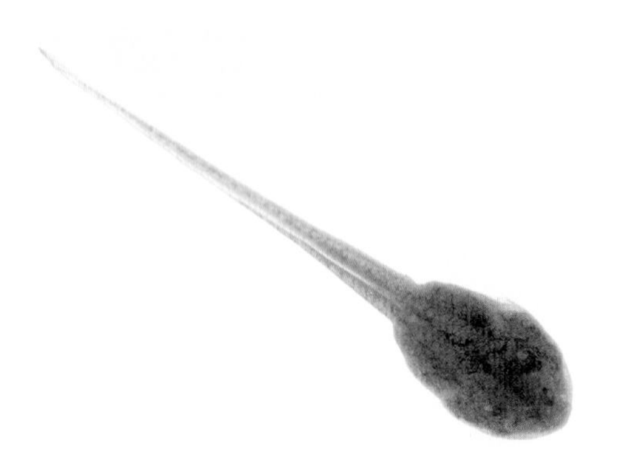

Stage 1: Tadpoles hatch from eggs. Tadpoles look more like fish than their parents. They have tails and live underwater. After a few weeks, each tadpole starts to change form.

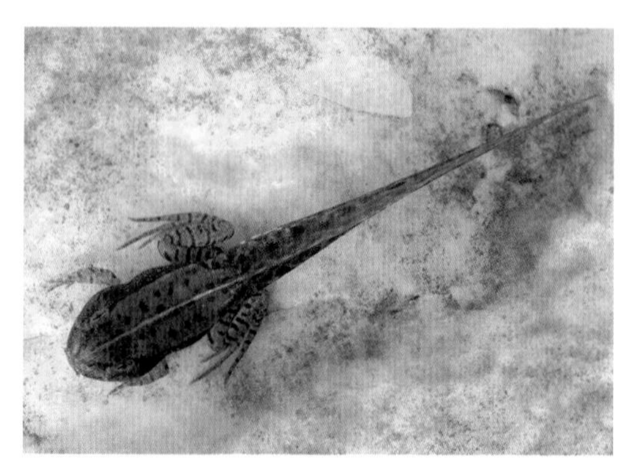

Stage 2: Soon, the tadpole starts to grow hind legs. Its body and tail start to shrink. The tadpole grows new parts to help it live on land.

Stage 3: Next, it grows front legs. The tail starts to disappear, and bones develop. The adolescent tadpole begins to use its lungs to breathe out of water.

Stage 4: Once they are fully grown adults, frogs can find mates and reproduce. Females lay eggs. At the end of their life cycle, frogs die, but their young continue the cycle.

Plants Have Life Cycles

Chapter

2

Strawberries grow in the wild. Farmers also plant strawberries for people to eat. Look at the two pictures of strawberry plants. How would you describe the difference between them?

Big Question

How do different types of plants change over their lifetimes?

The strawberry plant on the left does not have any fruit yet. The strawberry plant on the right has strawberries that are ready to pick and eat.

Strawberry plants are **flowering plants**. A flowering plant is any plant that has flowers. Flowers are part of the reproductive process of flowering plants.

Vocabulary

flowering plant, n. a plant that produces flowers during its life cycle

Plants have life cycles like animals do. The two plants in the pictures are in different life cycle stages.

Life Cycle of a Flowering Plant

When **seeds** have the right conditions, the young plant inside the seed gets bigger. Tiny roots and leaves grow out of the seed. When seeds begin to grow, it is called **germination**. Plant germination is somewhat like the birth of an animal. A new individual starts to grow.

The seed germinates.

Some plants live for only one year. Some live for many years and produce many seeds. A dying plant returns materials to the soil. Then the life cycle—germination, growth, reproduction, death—starts over!

The plant dies.

Fruit releases seeds.

All fruits contain seeds. Fruits often taste and smell good. Animals eat some kinds of fruits and spread the seeds. Other kinds of fruits fall from a plant onto the ground. Either way, the seeds end up in the ground.

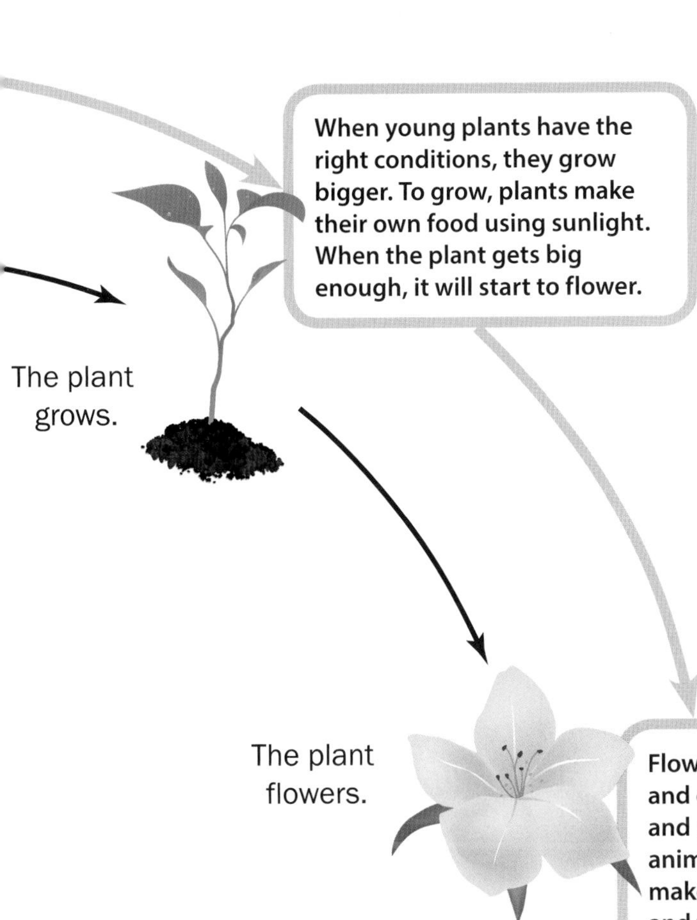

When young plants have the right conditions, they grow bigger. To grow, plants make their own food using sunlight. When the plant gets big enough, it will start to flower.

Vocabulary

seed, n. the part of a plant that protects the material that sprouts into a new plant

germination, n. the beginning of the growth process when a plant sprout comes out of a seed

pollination, n. the transfer of pollen that causes flowering plants to reproduce

The plant flowers.

Flowers come in many sizes, shapes, and colors. Many contain eggs and pollen. Flowers may attract animals that help move pollen to make contact with the eggs. Wind and water move pollen, too. The movement of pollen from one flower part to another is called **pollination**. After pollination, seeds begin to form.

The flower produces fruit.

When a flower is pollinated, a seed starts to grow. The flower has done its job! The petals start to drop off. The part of the plant around the seeds starts to grow and ripen. This is called the fruit. The fruit protects the seeds.

Life Cycle of a Fern

Some plants do not have flowers, fruits, or seeds. Ferns are nonflowering plants. They have been on Earth longer than the flowering plants. Ferns are often leafy. They usually grow in moist, shady places. Ferns have their own special kind of life cycle.

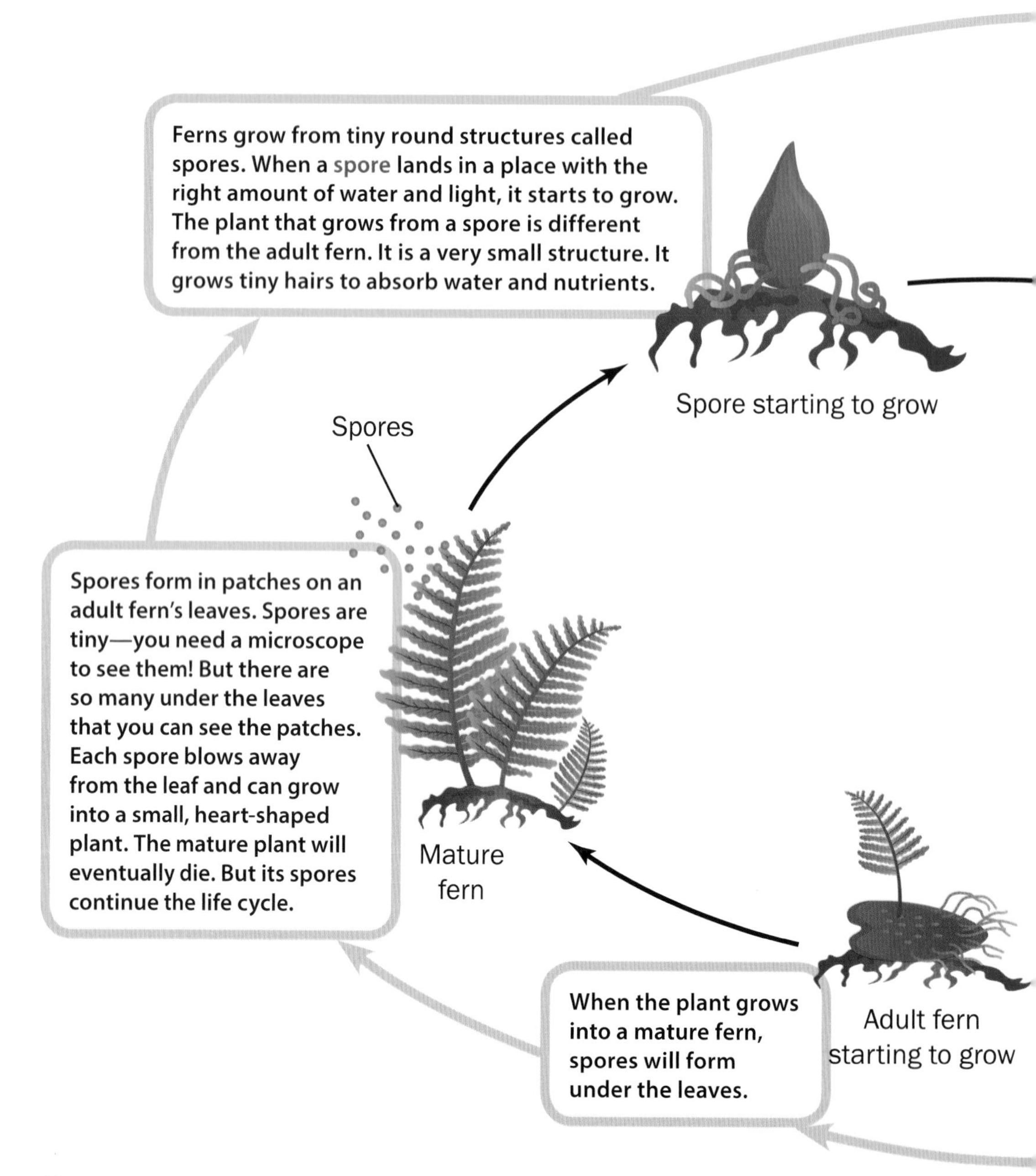

Vocabulary

spore, n. a tiny structure on some flowerless plants that can grow into a new plant

Spores do not need to be pollinated to grow into a new plant.

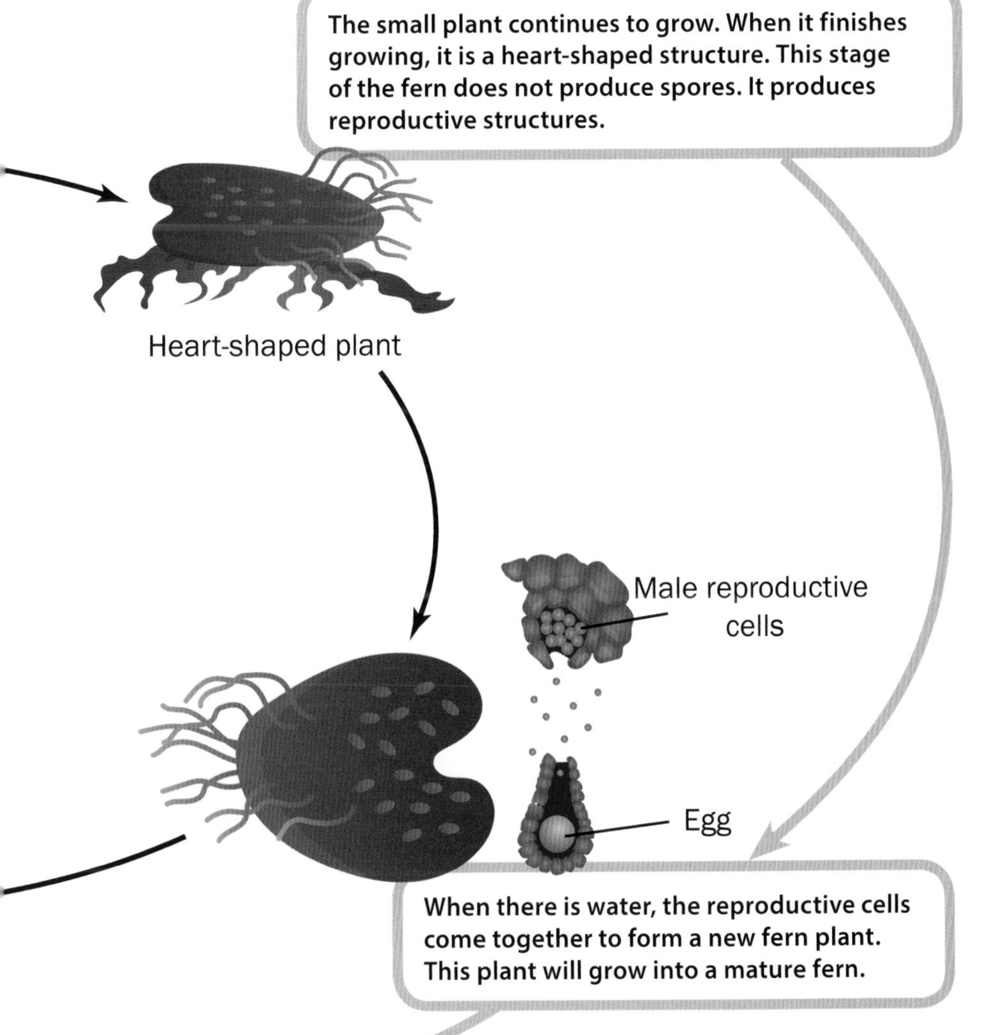

Ancient and Modern Plants

Ferns were on Earth before the flowering plants. Ferns have been on Earth for about 400 million years. The earliest evidence of flowering plants on Earth is from only 160 million years ago.

Scientists have found evidence of ancient ferns in fossils.

Flowering plants and ferns still exist today. Today's plants often look different from the kinds of plants that lived millions of years ago, but the patterns in their life cycles are similar. All plants start life, grow, reproduce, and die, just like animals.

Orchids may date back to the explosion of flowering plants 100 million years ago.

Traits of Parents, Offspring, and Siblings

Big Question

How are traits similar and different within families?

One thing is for sure: you can tell the students in your classroom apart. You all have similar features. You all have eyes, noses, and ears. But you and your classmates are not exactly the same. Features differ from one classmate to another.

This is true for other living things, too. All apple trees have trunks, branches, and leaves at some point in their life cycles. But all apple trees are not exactly the same. Any two apple trees can have many differences. They will likely have different numbers of branches. One might be taller than the other. Their trunks are probably not the same thickness.

Trees of the same type are not exactly the same.

Individuals Have Traits

An **individual** is one living thing, a single organism. The features used to describe an individual are called its **physical traits**. Living things of the same type, such as chickens, share similar traits and characteristics. But every individual has different traits that make it one of a kind.

Vocabulary

individual, n. a single living organism

physical trait, n. a feature of a living thing's body

How can you describe the traits of the chickens in the picture? Every chicken in the photograph has feathers and a comb on its head. Each has one beak and two featherless legs. And yet, no two chickens have the same feather pattern. These chickens differ. Their traits of feather color and pattern are not the same.

These chickens share similar traits, but they also differ.

Animal Offspring Inherit Traits from Their Parents

When plants and animals reproduce, the young are known as **offspring**. Offspring develop traits that are very similar to their parents as they grow. Look at the hippo parent and its offspring in the picture.

Hippo parents and their offspring have similar traits. Count their two nostrils, two eyes, and two ears just above the water's surface.

Vocabulary

offspring, n. a young organism produced by parents

inherit, v. to receive from parents

The two hippos in the picture look very much alike. This is because the young hippo received many traits from its two parents. When an organism receives traits from its parents, we say that the organism **inherits** the traits. Animal offspring inherit traits from their parents. Offspring resemble their parents, but they are not identical to their parents.

Black crowned crane offspring inherit long legs from their parents. The offspring and parent birds share similar traits. But the offspring do not look exactly like either parent.

Plants Inherit Traits from Their Parents

Gregor Mendel was an Austrian botanist. A botanist is a scientist who studies plants.

In the mid-1800s, a scientist named Gregor Mendel investigated how traits are inherited. He studied traits in pea plants. The traits included seed color, seed shape, flower color, and plant height. He noticed the traits are inherited in patterns from one generation of plants to the next.

For example, Mendel bred pea plants that only produced offspring with white flowers, generation after generation. He did the same with plants that only produced purple flowers. But when he bred white flowers with purple flowers, all of the new plants had purple flowers. However, when those offspring reproduced, a few of their offspring had white flowers! Mendel's experiments revealed patterns in how traits are passed on.

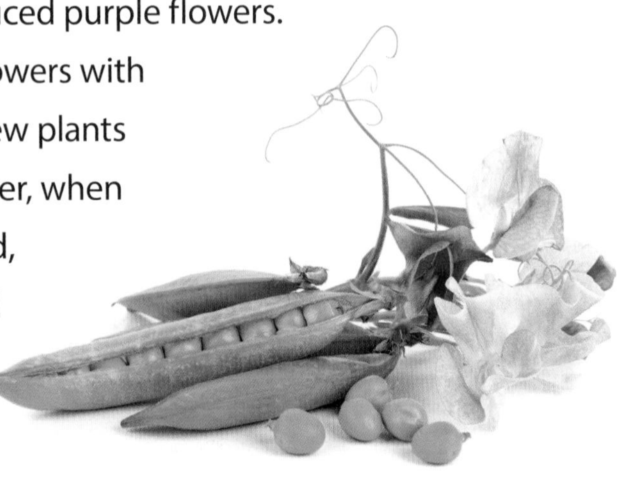

Each single pea is a seed of the pea plant. Each seed can grow into an offspring plant.

Other plants show patterns of inheritance similar to what Mendel found in peas. Offspring resemble their parents but are not identical to them. For example, these pictures show flowering plants called *four o'clocks*. They were given their name because they bloom around four o'clock. The flowers of different four o'clock plants may have different shades and colors.

Four o'clock flowers might look like the parent plants, or they might look much different. The offspring get traits from the parent plants, but they may not have identical traits.

If a plant with pink flowers reproduces with another plant with white flowers, the offspring can be red, white, or pink! Sometimes an offspring looks much different from its parents. The offspring flowers can even be striped or multicolored. Traits are inherited from parents in different patterns.

Siblings Inherit Similar Traits

Individuals inherit physical traits from their parents. A young parakeet might inherit its green feathers from its mother or inherit its orange beak from its father. But what if the same parents have more than one offspring?

This parakeet inherited its beak and feather colors from its parents.

Individuals that come from the same parents are called **siblings**. A sibling can be male or female, a brother or sister. Sometimes, two siblings inherit very similar traits and look very much alike. Some siblings inherit different traits and do not look similar at all. Do you know of any families in which all the children look exacty alike?

> **Vocabulary**
>
> **siblings, n.** organisms that come from the same parents

The litter of puppies in this picture appear very similar. But if you look closely, you will find differences between them. Each sibling inherited slightly different combinations of traits from the same parents.

These puppies are siblings. They look similar. But their fur color differs a little bit. Their faces have different shapes, too.

Siblings Are Rarely Identical

Most of the time, siblings just look similar. But sometimes, siblings can share **identical** inherited traits. Sometimes a mother has two or more babies in the same birth. In this case, the offspring are called multiples. Twins and triplets are multiples.

> **Vocabulary**
>
> **identical, adj.**
> exactly the same
>
> Identical twins inherit the exact same traits, causing them to look alike.

Some animals are very likely to give birth to multiples, such as mice. Mice and pigs often give birth to multiples. Identical siblings inherit identical traits from their parents. Their traits are identical when comparing one sibling to the other. Identical twins are rare among animals. Humans and armadillos are two types of animals that sometimes give birth to identical offspring.

These are nine-banded armadillos. Usually, when the armadillo mother has babies, she gives birth to four identical siblings. They are either all male or all female. Four babies born during the same birth are called *quadruplets*.

Siblings born at the same time can also inherit different traits. Sometimes twins or triplets can look so different that you cannot tell that they are related!

Traits Vary Within Families and Within Species

These ducklings have the same parents, but they do not look the same. One has yellow and white feathers. The other ducklings have darker feathers. Yet the ducklings are in the same family.

Traits vary within families. When traits **vary**, it means they show differences. Two parents and four offspring might have many traits in common. But members of the family also have differences between them. The ducklings look different because they have inherited different traits from their parents.

Vocabulary

vary, v. to change or differ

species, n. a group of similar individuals that can reproduce together

variation, n. a difference between things or in one thing over time

There are many different types of ducks that share similar traits. But wood ducks are different from domestic crested ducks. These ducks have different physical traits because they are different species. A **species** is a group of individuals that look very similar and that can reproduce together. The differences between two species are greater than **variations** of traits within a family.

A wood duck (left) and a domestic crested duck (right). These are both ducks, so they have similar traits. But they are different species, so some of their traits are very different.

Traits Vary Within Populations and Species

Chapter 4

Pete is a young emperor penguin. He is very similar to his sister, Polly. Pete and Polly inherited traits from their parents. The parents, Pete, and Polly are very much alike. But each of them is different in small ways. Traits vary among family members.

The penguin family lives on the Ross Ice Shelf in Antarctica. Many other emperor penguin families live there, too. All the penguin families that live there make up a **population** of emperor penguins.

All members of this population of emperor penguins look pretty much alike. Yet, as with families, members of this population vary in their traits. Traits vary within any population of living things.

Big Question

How do traits vary within groups of the same type of organism?

Vocabulary

population, n. a group of a single type of organism living in the same place at the same time

Many families make up a population of emperor penguins.

The penguins in Pete's population live with other populations in the same area. They live with populations of plankton, fish, squid, seals, algae, and orca. All the populations in this place interact with each other. To interact means to affect each other.

Word to Know

An *ecosystem* is all the living and nonliving things that interact in the same place.

All the living and nonliving things in an area make up an ecosystem. The penguin population is just one small part of the Ross Ice Shelf ecosystem.

Penguin populations and seal populations share the same ecosystem in Antarctica.

Traits Vary Within Populations

Katydids live in a meadow ecosystem. The katydid is a leaf-eating insect. Like all living things, any one katydid resembles its parents and siblings. It also resembles other katydids in the population it belongs to. A katydid might have a green body that looks like a leaf. But each katydid is different from all other katydids in its population. The trait of color can vary sharply within the katydid population. Some are pink! A pink katydid is not able to hide on a leaf as well as a green katydid can. The green katydid is less likely to be eaten, and the pink one is more likely to be eaten in a meadow with lots of green leaves. This means the green katydid may live long enough to reproduce. The offspring of the green katydid are more likely to be green, so they will be good hiders, too. But perhaps the pink katydid is more hidden when resting on a pink flower!

You can see that some traits can give an individual in a population a greater chance to survive than individuals with different traits. The ones who survive pass their traits on to offspring.

Pink katydids are more likely to be eaten when resting on a green leaf, so they are less likely to reproduce than green katydids.

Traits Vary Within Populations and Across Species

Here is another example of variation of traits in a population. The flamingo is a pink bird that lives in shallow water ecosystems. Flamingos have long, skinny legs and webbed feet. These traits help flamingos wade through water in search of their favorite food—shrimp! With webbed feet, they can walk on top of the soft mud without sinking and getting stuck.

Most adult flamingos in a population have about the same size feet. But some have smaller feet, and others have larger feet. Foot size, like the colors of katydids, is a trait that varies in a population.

A variation of a trait can help an individual in the population. For example, flamingos with larger webbed feet can stir up more mud. They may have a better chance of finding more shrimp! And that means they may have a better chance to survive and reproduce. But perhaps flamingos with small feet can fly away faster when they have to. This is another example of how some traits give some individuals an advantage in life.

Would feet the same shape as a flamingo's be an advantage to a hawk that needed to capture food with its feet?

More Trait Variation in Populations and Across Species

The saguaro is the largest cactus in the United States. It grows in hot, dry desert climates. Look at the picture of several saguaros. How do they vary?

These saguaros are the same species and around the same age, but they are different heights.

Saguaros have a way to survive seasons with little rain. They collect water through their roots in the soil when it rains. Then they store the water inside their green trunks. Saguaros' roots are shallow in the ground, only four to six inches deep. This lets them easily get water from the topsoil when it rains. The roots also reach far around the saguaro. The roots stretch out as far as the saguaro is tall.

Height is a trait that varies in saguaros. Saguaros grow very slowly. Over many years, they can grow up to sixty feet tall! Some saguaros can grow taller than others, even if they are the same age. Tall saguaros will have longer roots than shorter saguaros. Saguaros with longer roots can collect more water when it rains than saguaros with shorter roots. Saguaros that collect more water have a better chance of surviving in the desert.

Traits Vary Within Species

All the penguins in the world's many emperor penguin populations are part of the same species. A species is a group of living things that can reproduce together to make offspring that can also reproduce together. Human beings are all members of the same species.

You know that traits vary within families and within populations. Traits also vary among individuals of the same species. For example, the picture shows three frogs of the same species. Frogs of the same species all look similar and can reproduce together.

Wherever members of this species live, individuals with some traits have a better chance to survive and reproduce than other members of the species.

Three frogs of the same species have varied traits.

Environments and Traits

Chapter 5

Flamingos are wading birds. These two are the same species. But they come from different **environments**. How can you tell?

Big Question

How does the environment affect the traits of living things?

Vocabulary

environment, n. a surrounding area that contains living and nonliving things

The color of the flamingos' feathers is a trait that can change depending on what they eat.

One flamingo has feathers that are bright pink. The other has feathers that are pale pink. These flamingos have been getting their food from different places. Their colors are different because of what the flamingos eat!

Some environments contain more shrimp than others. The flamingos that live in these areas will be darker pink than other flamingos. This is how the environment can affect the color of a flamingo's feathers. Some traits can vary because of where an individual lives. Flamingo color traits are not completely inherited.

A Living Thing's Weight Can Be Affected by the Environment

There are many ways the environment can influence the traits of living things.

Look at the cat in the picture. This indoor cat is fed a lot of food!

The cat has certain traits that won't change much, such as its fur color and eye color. But the cat's weight is a trait that varies because of its environment. This trait of being this large is not inherited.

Any animal that is given too much food and has too little exercise can become overweight.

Since this cat lives indoors, it does not hunt for its own food. It also does not jump, climb, or run as much as it would outdoors. The cat's owner provides more food than the cat needs. As a result, the cat has a high body weight. The cat's weight is a trait that varies because of its environment.

A macaque /muh*kak/ is a species of monkey. Its natural environment is the jungles of Asia. Many macaque populations have found their way into human cities, though.

These macaques are heavier than macaques that live in wild jungle areas.

Monkeys in a jungle are very active moving from tree to tree. But the city is a new environment for monkeys. Now, they can find and eat foods that people eat. These foods are not healthy for them. Monkeys that live in the city tend to weigh more than monkeys that live in the wild.

A Living Thing's Behavior Can Be Affected by the Environment

You know that the environment can affect a living thing's physical traits. An animal's behavior also can be affected by the environment. Behavior is how animals act. Behavior patterns are traits. Some are inherited, and some can be the result of the environment.

Meerkats burrow into the ground. Their burrows help them stay safe from predators and stay cool in hot environments.

The meerkat is a small animal that lives in the deserts of southern Africa. Meerkats live in large groups. By living in a group environment, meerkats first learn and then teach other meerkats behaviors that help them survive.

Certain meerkats look out for danger while others in the group search for food. Some individuals stand guard to keep groups safe. They look up into the sky for dangerous birds. If they see a predator bird, they make a warning sound. This behavioral trait keeps the food hunter meerkats safe.

The prairie dog is another animal that has varying behavioral traits depending on the conditions of the environment. Prairie dogs are small rodents that burrow underground. Like meerkats, prairie dogs learn from living in large groups.

Prairie dogs live with many groups of families underground.

Prairie dogs first learn and then teach each other about what to eat and where to find food.

Living in groups also helps young prairie dogs learn how to make burrows with connected tunnels. Burrowing is a skill that helps them survive.

Prairie dogs use tunnel systems for shelter and survival.

Environment Can Affect a Plant's Traits

Plants need water to grow and survive. In the picture of potted plants, can you tell which plant got enough water and which one did not?

Sometimes environments have dry seasons with very little water, or they have rainy seasons with a lot of water. When an environment does not have enough water, it changes the traits of the plants that live there. The plants might dry out. They can turn brown, lose their leaves, and stop growing. They will no longer be as green and healthy as they once were.

Too much water is not good for plants either! An environment can get too much water during a season that is very rainy. Even though plants need water to live, too much water can make plants die.

When the amount of water in an environment changes, it can change the physical traits of the plants in that area.

When too much water stays in certain areas, plant roots may not get enough oxygen.

Helpful Traits

Chapter

6

Remember katydids? It is harder for predators to spot a green katydid on a green leaf. It is easier to spot a pink one on a green leaf. So, pink katydids on green leaves are more easily eaten than green ones. The green katydid has a trait that is an advantage. An **advantage** is something that is helpful to the organism.

Look at the two foxes. Which one do you think will have an easier time sneaking up on prey in the snowy north? The fox that catches more food will have a better chance of surviving the winter.

Big Question

Which traits are helpful to organisms, and which traits are not?

Vocabulary

advantage, n. a factor that helps an organism

survive, v. to stay alive

When an organism has an advantage over other members of its population, it is more likely to survive and reproduce. Organisms that **survive** live to become adults, and adults reproduce.

The fox that blends into its environment is more likely to catch prey and survive.

Helpful Traits

The large ground finch is a bird with an unusually large beak. Charles Darwin was a famous scientist who first observed this species in the Galápagos Islands. Scientists knew that this bird was very similar to other finches living on nearby islands. But this one had a larger beak. Darwin wondered if this large beak gave the finch an advantage in its environment.

Darwin's observations showed that there were a lot of plants with very hard seeds in this finch's environment. The large ground finch was able to crack the hard seeds because it had such a large, strong beak.

Darwin discovered other finches in the Galápagos Islands. He found the warbler finch with a much smaller pointed beak. This beak shape helps the warbler finch catch insects in small holes. Darwin saw a woodpecker finch with a longer beak. This beak shape helps the woodpecker finch use twigs to catch grubs and worms. Different species of finches had different kinds of beaks. But each species had an advantage over other finches in its environment.

The large ground finch has a thick, strong beak that helps it crack hard seeds.

Helpful or Not? It Depends on the Environment!

A physical trait can either be helpful or harmful. It depends on the organism's environment. The prairie dog digs burrows in the ground. The burrow has a small opening that predators can't fit through easily.

A prairie dog born with long legs will probably run faster aboveground than its siblings. This would be helpful for escaping predators. But the prairie dog hides from predators in tiny burrows. If its legs are longer, it may not move well in tight burrow spaces. The burrow for a prairie dog with longer legs would need to have a larger opening. A predator could get into that burrow more easily. So, a prairie dog with longer legs might be more easily caught by a predator.

Some Traits Help Reproduction

The green color of a katydid can help it blend in. A thick beak can help a finch crack seeds. But some traits exist to help an animal attract a mate.

If an animal attracts a mate, it is more likely to reproduce. Colorful skin patterns or feathers are some of the traits that can help animals attract mates.

The Inca tern is a bird that grows long feathery tufts on its head. The tufts look like a white mustache. This colorful decoration tells other terns that it is healthy and a good choice for a mate. An Inca tern with a long, bright mustache will attract more mates than a tern with a smaller mustache. This helpful trait makes it more likely that the tern will reproduce. The tern will pass this trait on to its offspring.

The white mustache of the Inca tern helps it attract a mate.

Helpful Traits in Plants

Let's not forget plants! Plants also have many physical traits to help them survive. Some traits help plants survive in harsh environments. Other traits help protect plants from animals that might eat them. Some traits help plants reproduce by attracting pollinators.

The Mojave Desert is a very dry environment. Plants and animals have little water to survive. Many animals get their water by eating plants. For desert plants to survive, they must have traits that protect them from thirsty animals.

The cholla cactus has an obvious trait that gives it protection. The plant is covered in sharp spines. The spines are dense and long. Few animals can reach the plant parts that hold water.

Some cholla cacti have more spines than others. A cholla cactus with more spines is less likely to be eaten. If it survives long enough, the plant will produce flowers and reproduce.

Desert plants like the cholla cactus use sharp spines to protect them from thirsty animals.

A Trait Can Be a Disadvantage

Some animals are born without color in their skin or hair. They are usually white all over and have pink eyes. These animals are called *albinos*. In some situations, this really might help the individual.

But it can also be a trait that is harmful. With white fur or skin, it is hard for an animal to hide from predators. If an albino is a predator, it is hard for the animal to sneak up on its prey. Also, albinos often do not see as well as other members of their population. These traits can make it harder for an albino to survive. This is an example of a harmful trait in animals.

Look at the albino alligator in the picture. The alligator is easy to see against the dark background. Birds, mammals, and other prey can easily see the alligator, too. This tells them to stay away. It will be hard for this alligator to get food. This alligator's white color is not an advantage.

Albino animals like this alligator have a hard time hiding in their environment.

Traits, Survival, and Differing Environments

All living things can be described by their traits—hairy bodies, long ears, fast running, or spotted colors. Traits usually help an organism survive in its environment. But if the environment changes, a helpful trait may no longer help.

Big Question

Why are traits important when environments change?

In the open grasslands of Africa, giraffes have room to move their large bodies around. They can also see long distances to watch for predators. In the grasslands, a giraffe's long legs and long neck are helpful traits. But what if the giraffe's environment changes? What if the open grasslands became thick, tropical forests because it started to rain more and more over a long time? The giraffe would not survive as well then. Its long neck and long legs would get tangled in the tree branches and vines. It would be harder for the giraffe to see predators. The giraffe's traits that were helpful in the grasslands would no longer be helpful in the tropical forest.

The long legs and long necks of giraffes are helpful traits in the open grasslands. In thick forests, those same traits may threaten survival.

Environmental Changes Affect Survival

Air Pollution and Peppered Moths

Peppered moths have white wings and tiny black specks. During the day, the moth rests on tree trunks. The trees are light colored. The peppered moth blends in so well that birds and other predators cannot see it. These moths survive because they are **camouflaged**. Moths that are camouflaged are hidden.

At first, it looks like there is only one black moth in this picture. Look closer. There is a second moth!

Some peppered moths are born with black wings. Since the tree trunks are light colored, the dark peppered moths are easily seen by birds. The dark moths are eaten more than the light moths. Few of them survive long enough to reproduce and pass on their physical traits. Because of this, there are fewer members of the population with dark wings.

Vocabulary

camouflage, n.
a color or pattern that helps an organism blend in with its surroundings

Scientists have debated about this moth for a long time. Some, but not all, scientists suggest a possibility about them.

In the 1900s, humans began using coal to power factories. Burned coal makes black powder in the air called soot. Trees became covered in black soot. The peppered moth's environment had changed. In this changed environment the light form of the peppered moth was much easier for birds to see against the dark tree bark. The light moth was no longer camouflaged.

Birds ate more and more light moths. Black moths had better camouflage. They survived to make more offspring. Soon the dark moth was common, and the white moth was **rare**. The change in the peppered moth's environment caused a helpful trait to become harmful!

> **Vocabulary**
>
> **rare, adj.** not found in large numbers

Today, factories do not make as much soot. The tree trunks have returned to their normal light color. The light form is once again a helpful trait. It is more common than the dark form.

Pollution from coal factories changed the habitat of the peppered moth. Light-colored moths were no longer able to hide on the soot-covered trees.

Climate and Elephants

Today, elephants live mostly in parts of Africa and Asia. These environments are warm. They provide plenty of vegetation to eat. But what if the climate changed over time? What if it grew much colder? Most elephants would not survive. They do not have the traits to survive in cold, snowy climates.

Yet, close relatives of elephants did once live in such climates. How? They had long, thick fur and smaller ears. The ears of modern elephants are big to help them stay cool. Blood vessels circulate through the ears, carrying heat. When the elephants fan their ears in the breeze, they lose some of that heat.

If the climate changed today, would elephants survive? What if a baby elephant happened to be born with thicker, longer hair or smaller ears? Would it be more likely to survive? It could pass those traits on to its offspring. Over time, its descendants would be better adapted for the cold and snow than other kinds of elephants.

In the past, woolly mammoths survived harsh climates because of long hair, extra fat, and smaller ears.

Human Explorers and Dodo Birds

Living things themselves can cause environmental changes. The dodo was a species of bird that lived on a single island in the Indian Ocean. The dodo was adapted to living on the ground and could not fly. It ate fruits that fell from trees. Adult dodos made their nests on the ground. Before humans came to the dodo's island habitat, the dodo had no predators.

Explorers settled on the island in the 1600s. Humans were a new species that entered the environment and caused it to change. They began hunting the dodo and taking dodo eggs from nests for food. The dodos could not fly away. They had no traits that helped the species live in an environment with human predators. Eventually dodos became **extinct**.

Vocabulary

extinct, adj. having no remaining living members

Sometimes a species cannot adapt to environmental change when a new species arrives.

Melting Ice and Polar Bears

Polar bears live in the Arctic Circle. Their habitat is the land and ocean that cover the northern parts of Earth. Polar bears spend most of their time on large floating ice sheets. They hunt for seals on the ice sheets. Polar bears must capture many seals to survive the cold winters.

Earth's climate is getting warmer, and the Arctic ice sheets are melting. The ice sheets today have less surface area than they did thirty years ago. Many of the ice sheets are smaller and separated by open water. This means that polar bears must swim long distances to find habitats for hunting seals. Many polar bears are not able to find places to hunt. Without food, they cannot survive.

Remember: all living things are fit to live in their environment. They have traits that ensure survival. But when the environment changes, only those individuals with certain traits will survive and reproduce. If not enough individuals have the right traits for survival, the species will become extinct.

Polar bears' traits are not adapted to living with much less ice surface.

Glossary

A

adolescence, n. a stage of the life cycle when a young animal is developing into an adult (2)

advantage, n. a factor that helps an organism (33)

C

camouflage, n. a color or pattern that helps an organism blend in with its surroundings (40)

E

environment, n. a surrounding area that contains living and nonliving things (27)

extinct, adj. having no remaining living members (43)

F

flowering plant, n. a plant that produces flowers during its life cycle (7)

G

germination, n. the beginning of the growth process when a plant sprout comes out of a seed (9)

I

identical, adj. exactly the same (19)

individual, n. a single living organism (14)

inherit, v. to receive from parents (15)

L

life cycle, n. a series of stages in an organism's life (1)

M

metamorphosis, n. a change of form during the life cycle of some animals (5)

O

offspring, n. a young organism produced by parents (15)

organism, n. any living thing (1)

P

physical trait, n. a feature of a living thing's body (14)

pollination, n. the transfer of pollen that causes flowering plants to reproduce (9)

population, n. a group of a single type of organism living in the same place at the same time (21)

R

rare, adj. not found in large numbers (41)

reproduction, n. the process of making new organisms (3)

S

seed, n. the part of a plant that protects the material that sprouts into a new plant (9)

siblings, n. organisms that come from the same parents (18)

species, n. a group of similar individuals that can reproduce together (20)

spore, n. a tiny structure on some nonflowering plants that can grow into a new plant (11)

survive, v. to stay alive (33)

V

variation, n. a difference between things or in one thing over time (20)

vary, v. to change or differ (20)

CKSci™

Core Knowledge Science™

Series Editor-in-Chief
E.D. Hirsch Jr.

Editorial Directors
Daniel H. Franck and Richard B. Talbot

Subject Matter Expert

Joyce Latimer, PhD
Professor
Department of Horticulture
Virginia Tech
Blacksburg, Virginia

Illustrations and Photo Credits

Age fotostock / Alamy Stock Photo: Cover C, 16a
Antarctica / Alamy Stock Photo: 22
Anton Sorokin / Alamy Stock Photo: 23a
Ardea.com / Mary Evans / Tonci / Pantheon / SuperStock: 5c
Bill CosterPanthe / Pantheon / SuperStock: 40
Biosphoto / SuperStock: 4d
Clarence Holmes / age fotostock / SuperStock: 5d
Cropper / Alamy Stock Photo: 24a
Cultura Limited / Cultura Limited / SuperStock: 2a, 32a
Danuta Hyniewska / age fotostock / SuperStock: 20b
Don Johnston / age fotostock / SuperStock: 6e
Duncan Usher / Pantheon / SuperStock: 35
Elena Elisseeva / Alamy Stock Photo: Cover A, 16b
Emi Allen / SuperStock: 31a
Eric Baccega / age fotostock / SuperStock: 39
Felix Choo / Alamy Stock Photo: 14
FoodCollection / SuperStock: 7b
Frank Siteman / age fotostock / SuperStock: 18b
Fredo de Luna / fStop Images / SuperStock: 2c
Gina Kelly / Alamy Stock Photo: 24b
Ian Cook / All Canada Photos / SuperStock: 37
ImageBROKER / SuperStock: 2b, 6c, 17b, 20a
Ivan Mochtar Olianto / Alamy Stock Photo: 29
Jean Michel Labat / Pantheon / SuperStock: 4c
Jennifer Booher / Alamy Stock Photo: 20c
John Daniels / Pantheon / SuperStock: 18a
Juniors / SuperStock: 7a, 15b, 17a, 17c, 28, 30
Kanwarjit Singh Boparai, DVM / Alamy Stock Photo: 36
Leena Robinson / Alamy Stock Photo: 19
Martin Shields / Alamy Stock Photo: 12a
Matthias Breiter / Minden Pictures / SuperStock: 33
Moodboard / SuperStock: Cover D, 26
National Geographic Image Collection / Alamy Stock Photo: 38
Naturbild / Johner / SuperStock: 4b, 6b
NaturePL / SuperStock: i, iii, 5e, 15a
NHPA / SuperStock: 1, 6d
Paul Roget / Ikon Images / SuperStock: 43
PhotoAlto / SuperStock: 32b
Radius / SuperStock: 21
Roberta Canu / Alamy Stock Photo: 5b
RubberBall / SuperStock: 13
Science Photo Library / Alamy Stock Photo: 23b
Science Photo Library / SuperStock: 12b
Science Picture Co / SuperStock: 42
Tim Fitzharris / Minden Pictures / SuperStock: 25
Westend61 / SuperStock: 27
Wolfgang Kaehler / SuperStock: 34, 44
Zen Shui / SuperStock: 4e

●儿童记忆潜能开发丛书●

中国人口出版社
China Population Publishing House
全国百佳出版单位

图书在版编目（CIP）数据

记忆力训练 : 全6册 / 真果果主编. -- 北京 : 中国人口出版社, 2014.1
（儿童记忆潜能开发丛书）
ISBN 978-7-5101-2170-8

Ⅰ. ①记… Ⅱ. ①真… Ⅲ. ①记忆能力－能力培养－学前教育－教学参考资料 Ⅳ. ①G613

中国版本图书馆CIP数据核字(2013)第294947号

记忆力训练2
真果果 主编

出版发行	中国人口出版社
印　　刷	北京卡乐富印刷有限公司
开　　本	787毫米×1092毫米　1/16
印　　张	18
版　　次	2014年1月第1版
印　　次	2014年1月第1次印刷
书　　号	ISBN 978-7-5101-2170-8
定　　价	75.00元（全6册）

社　　长	陶庆军
网　　址	www.rkcbs.net
电子信箱	rkcbs@126.com
总编室电话	（010）83519392
发行部电话	（010）83514662
地　　址	北京市西城区广安门南街80号中加大厦
邮　　编	100054

目 录

场景记忆

图形配对

曼陀罗涂色

记忆力游戏精选

纸牌排队游戏是一种简单、易操作的记忆力游戏，锻炼孩子的方位记忆能力，拓展他们的逻辑思维空间。家长可根据孩子自身的能力和对游戏的掌握程度，逐渐增加游戏的难度。常见的排列方式有AB式、ABC式、ABB式、ABA式等。下面，我们就以最简单的AB式为例，讲解一下游戏规则。

首先，按照猫–兔–猫–兔–猫–兔的顺序，排好6张牌，让孩子记忆。

然后，打乱纸牌顺序，让孩子根据记忆，排出正确顺序。

提示：要照顾孩子的积极性，第一次游戏时适当降低难度，只打乱后4张纸牌的顺序。当孩子熟练掌握游戏规则后，再增加难度。纸牌张数可递次增加，排列方式也可自由变化。

小动物与水果

仔细观察下图，记忆一下，然后翻到下一页，将不在本图中的小动物或水果圈出来。

游戏难度 ★★★☆☆

哪个小动物或水果在上页没有出现过?

场景记忆

小熊的花

仔细观察下图，记忆一下，然后翻到下一页，将不在本图中的花圈出来。

游戏难度 ★★☆☆☆

哪束花在上页没有出现过?

野餐

仔细观察草地上的食物，记忆一下，然后翻到下一页，将不在本图中的食物圈出来。

游戏难度 ★★★★☆

哪两种食物在上页没有出现过?

体育比赛

仔细观察下图，记忆一下，然后翻到下一页，将本图中出现的小动物圈出来。

游戏难度 ★★★★☆

哪些小动物在上页出现过？要仔细辨别它们的动作哟！

场景记忆

医疗用品

仔细观察图中的医疗用品，记忆一下，然后翻到下一页，将不在本图中的物品圈出来。

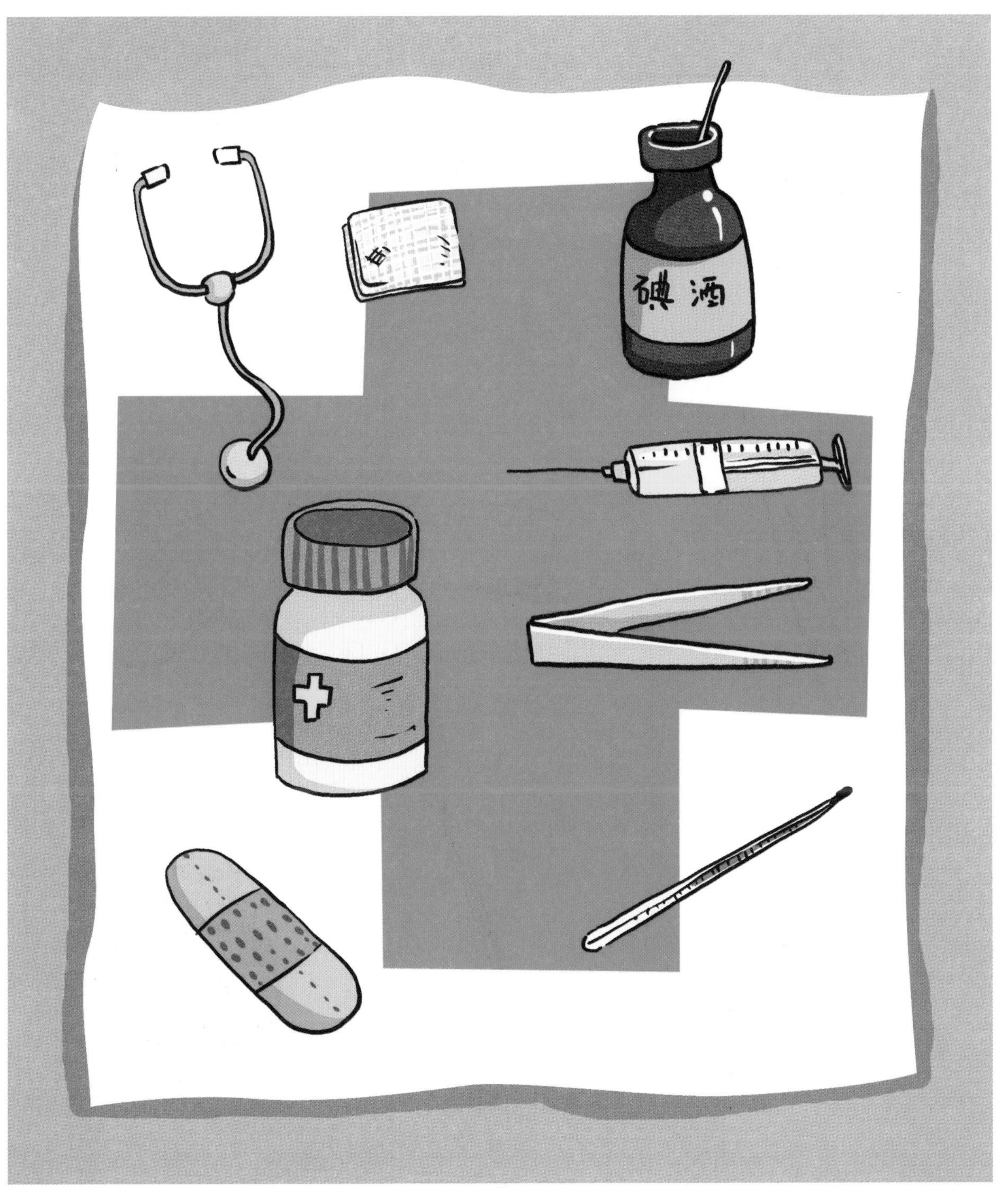

游戏难度 ★★☆☆☆

哪个物品在上页没有出现过？

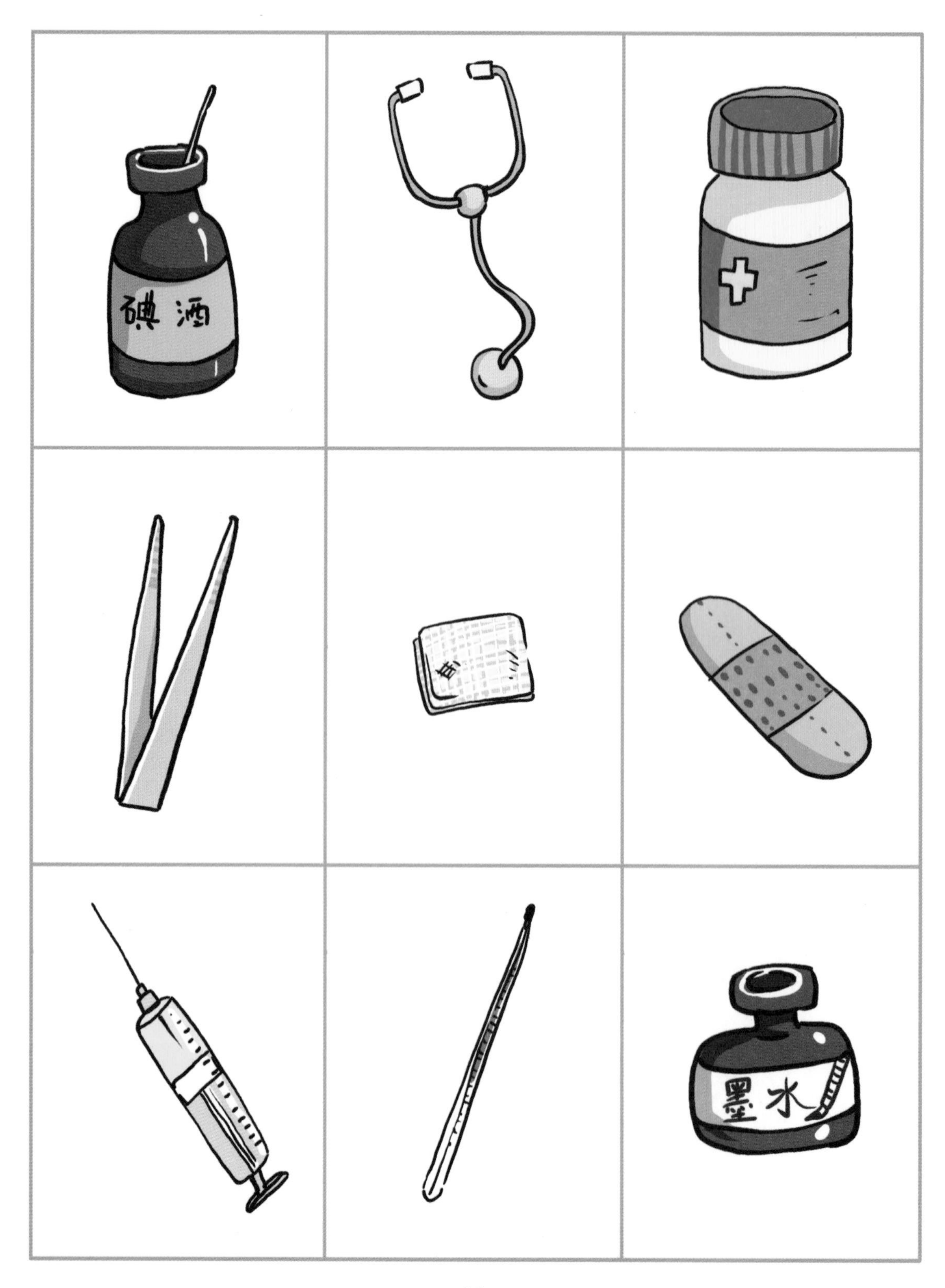

动物表情

仔细观察下图，记忆一下，然后翻到下一页，将不在本图中的3个表情圈出来。

游戏难度 ★★★☆☆

哪3个表情没在上页出现过?

草地上的垃圾

仔细观察下图，记忆一下，然后翻到下一页，将本图中出现的垃圾圈出来。

游戏难度 ★★★☆☆

哪个垃圾在上页出现过?

垃圾分类

仔细观察下面2组图，记忆一下，然后翻到下一页，将垃圾与对应的垃圾箱连线。

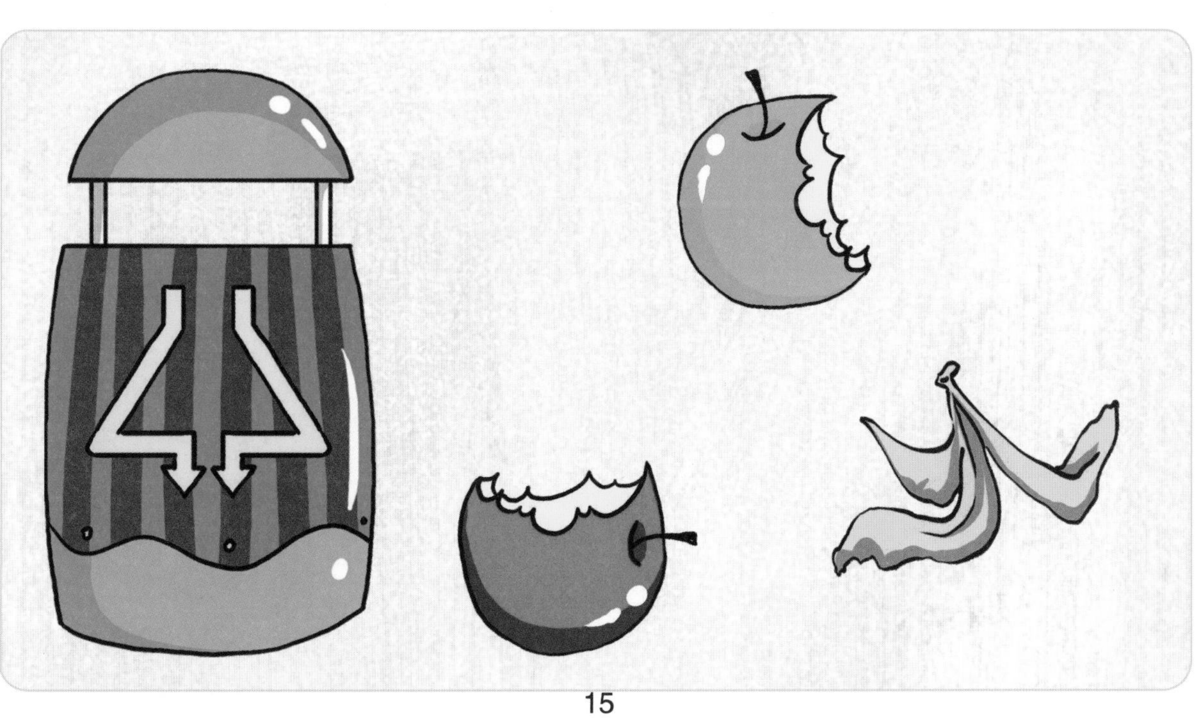

游戏难度 ★★★☆☆

将同一组的垃圾箱与垃圾连起来。

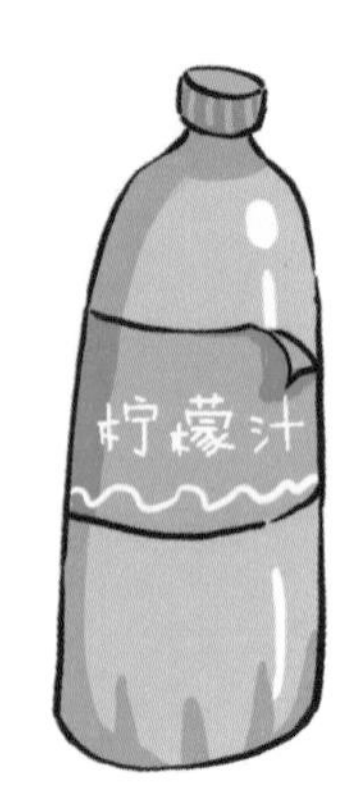

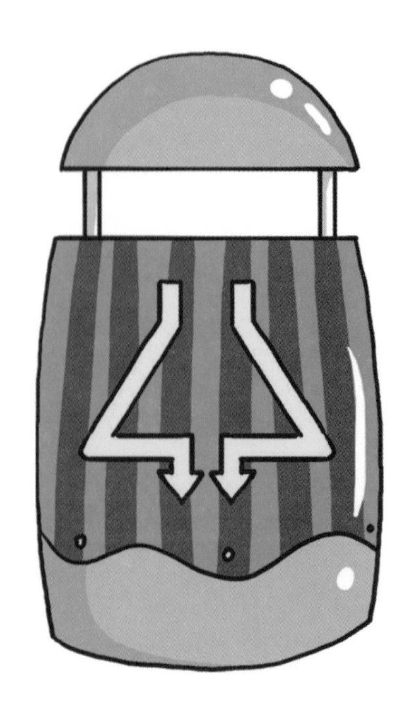

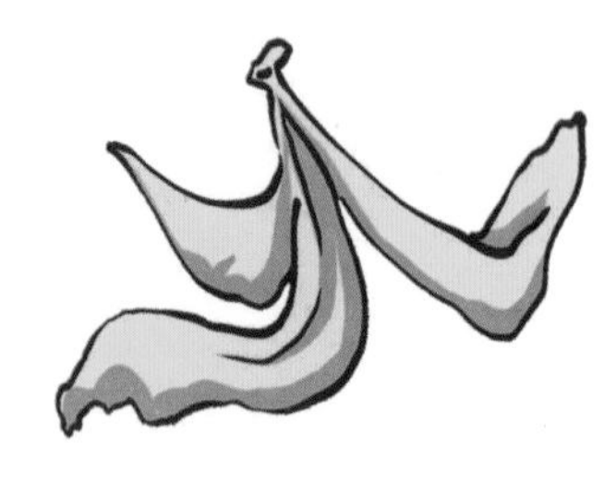

垃圾换鲜花

仔细观察下面3组图，记忆一下，然后翻到下一页，将同一组的物品连线。

游戏难度 ★★☆☆☆

将同一组的物品连起来。

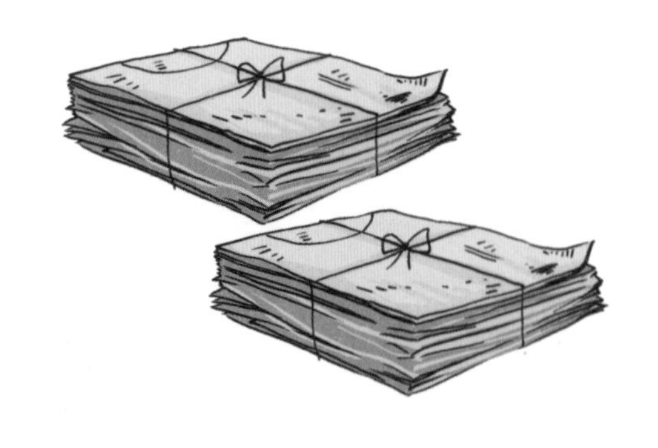

植树

仔细观察下图，记忆一下，然后翻到下一页，将小动物和它们所植的树连线。

游戏难度 ★★★★☆

将同一组的动物与树连起来。

小鸟与大树

仔细观察下图，记忆一下，然后翻到下一页，将大树与对应的小鸟连线。

游戏难度 ★★★★☆

将同一组的大树与小鸟连起来。

鱼缸和鱼

仔细观察下图，记忆一下，然后翻到下一页，将鱼缸和对应的鱼连线。

游戏难度 ★★★☆☆

将同一组的鱼缸和鱼连起来。

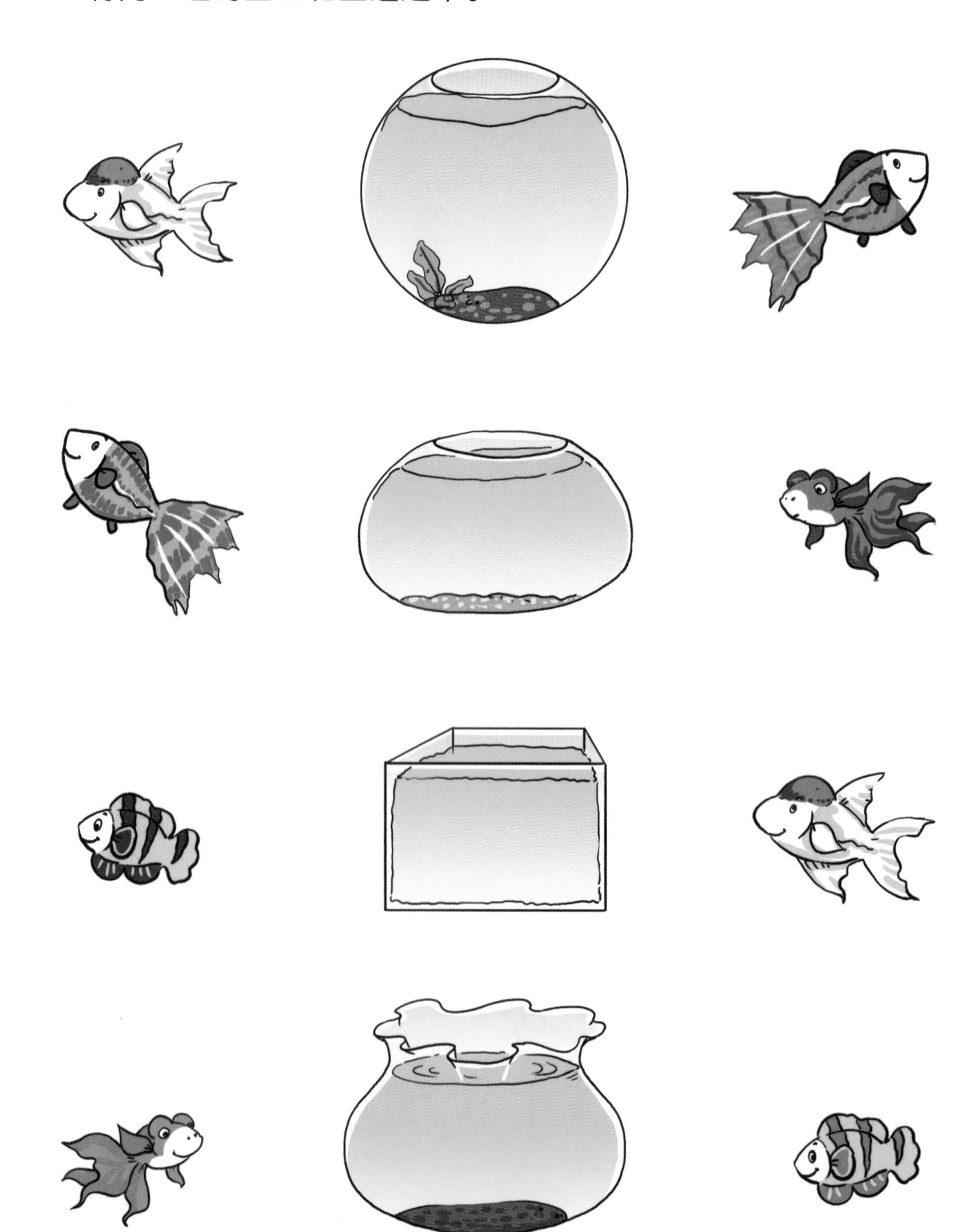

数字风铃

仔细观察下图，记忆一下，然后翻到下一页，将小动物头像与数字连线。

游戏难度 ★★★★☆

将同一组的小动物头像与数字连起来。

曼陀罗涂色（一）

看下面的曼陀罗图案，记住每一格中的颜色，翻到下一页，填涂上去。

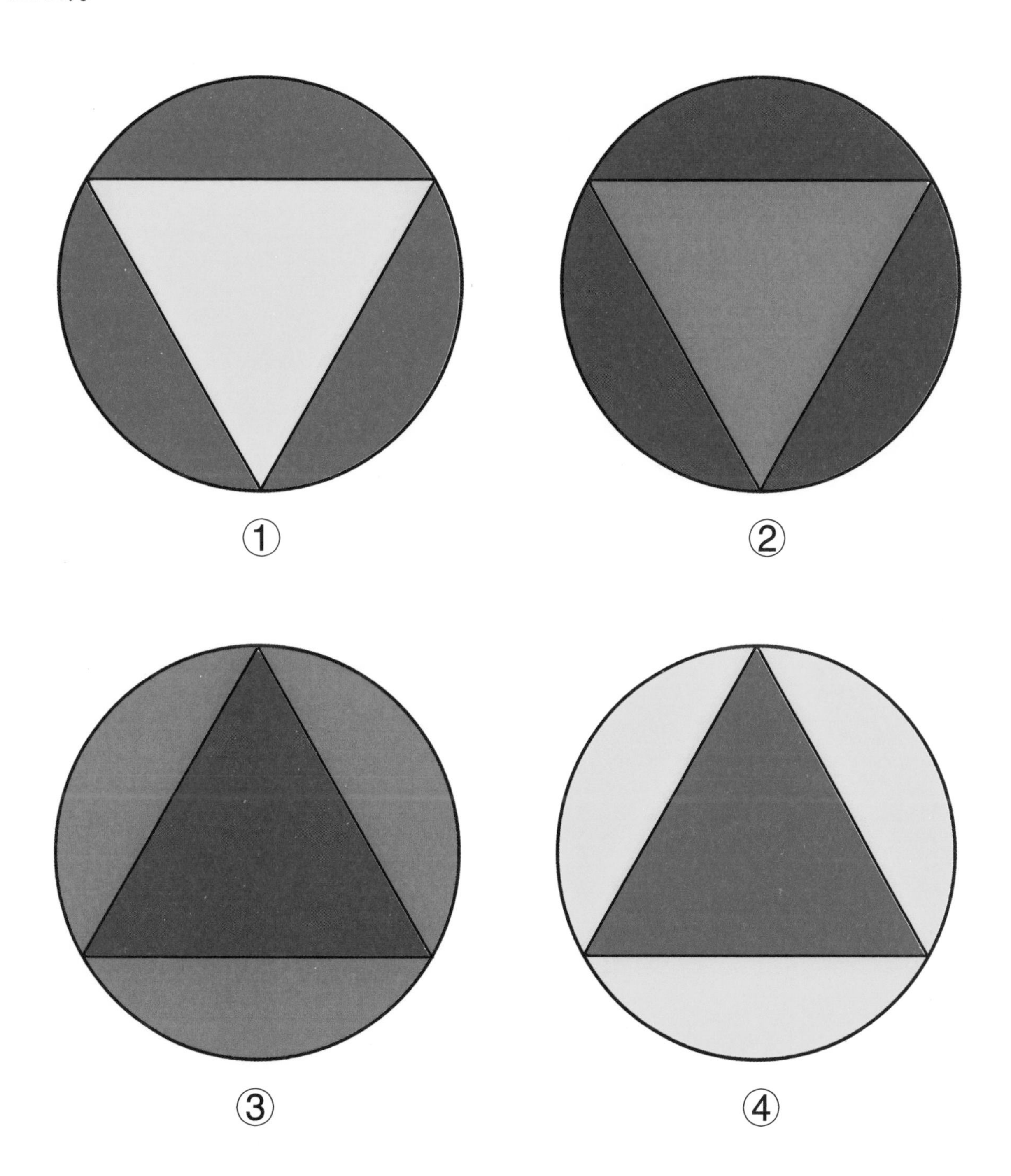

游戏难度 ★★☆☆☆

给下列曼陀罗图案涂上颜色。

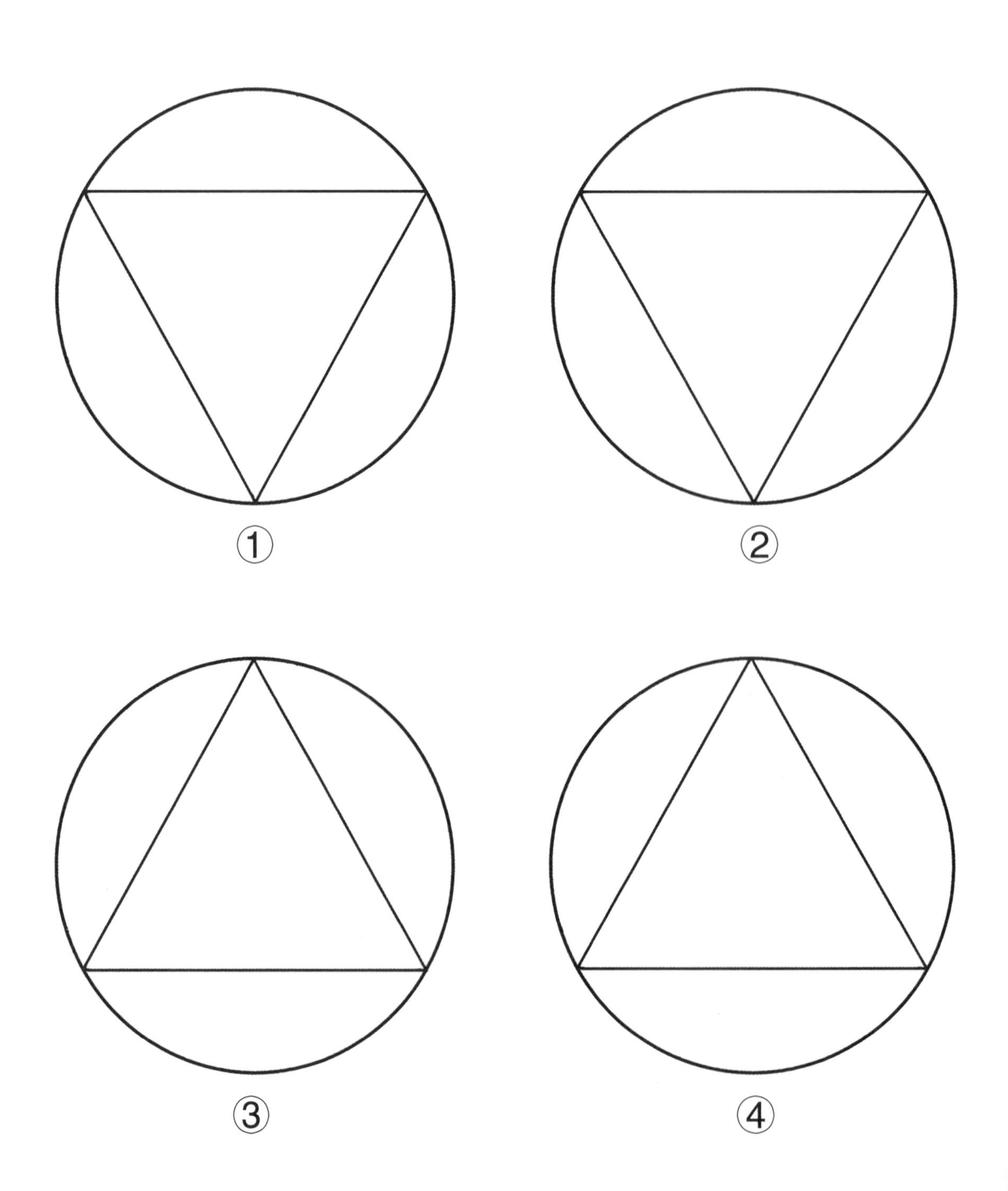

曼陀罗涂色（二）

看下面的曼陀罗图案，记住每一格中的颜色，翻到下一页，填涂上去。

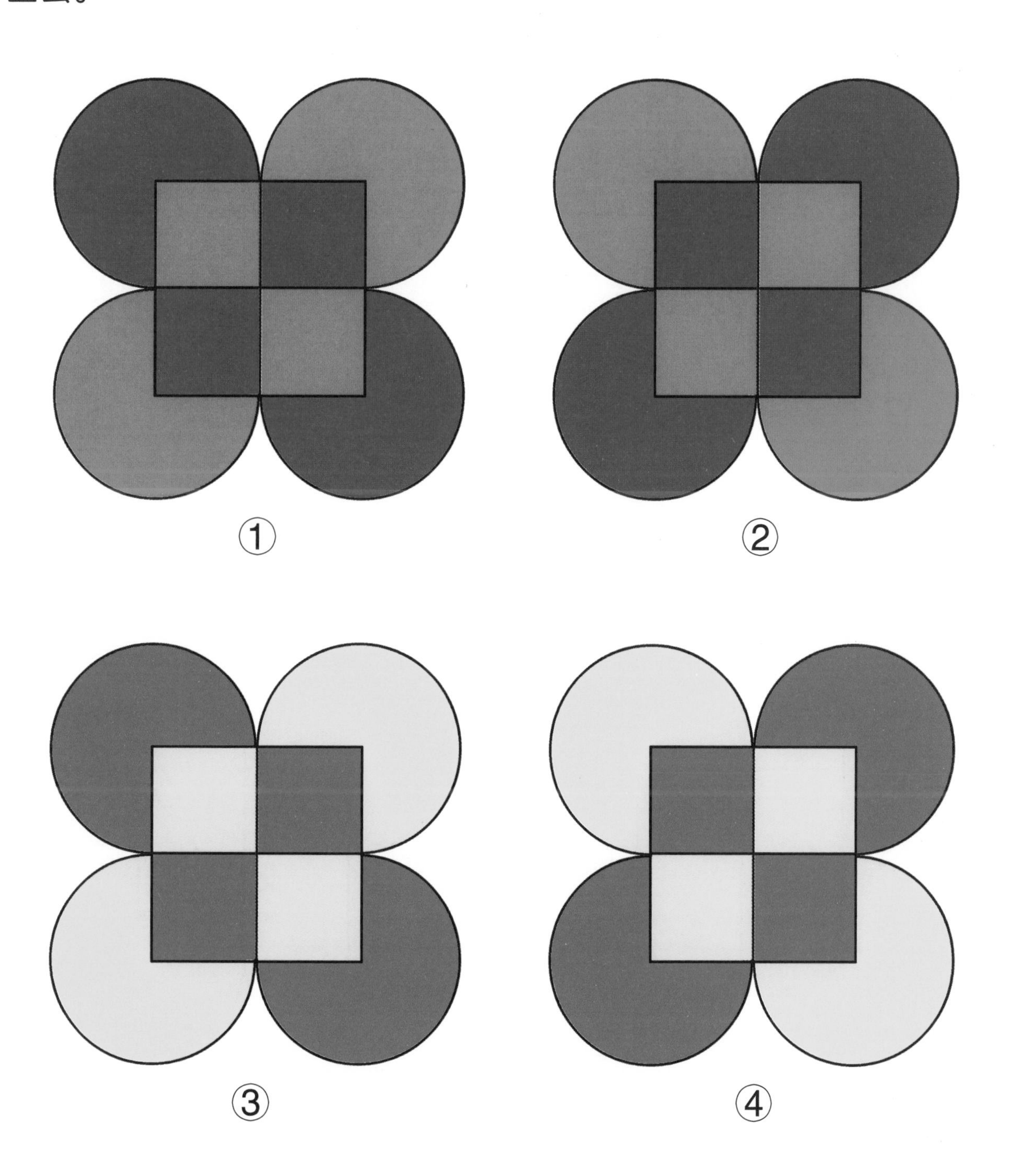

游戏难度 ★★★☆☆

给下列曼陀罗图案涂上颜色。

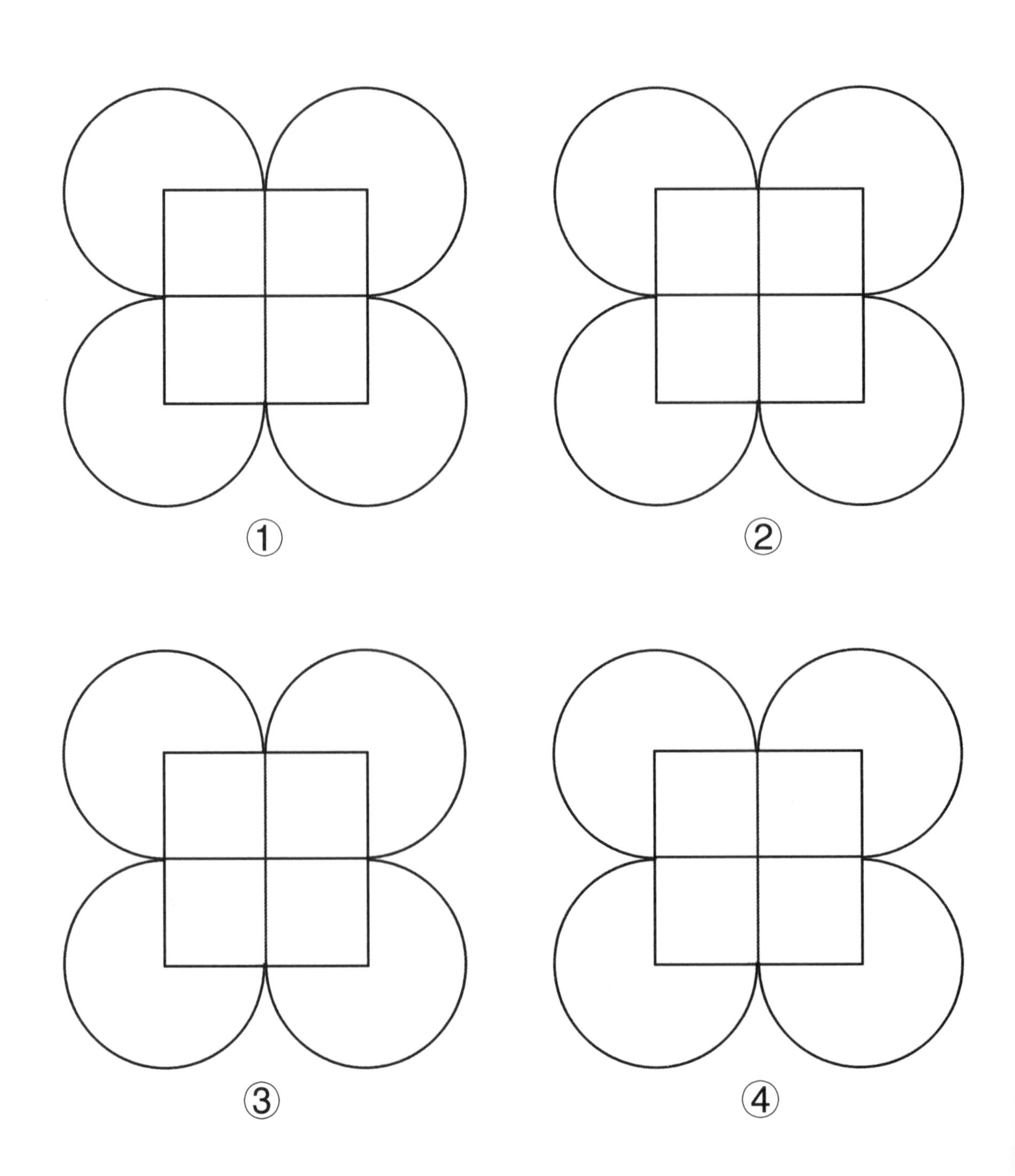

曼陀罗涂色（三）

看下面的曼陀罗图案，记住每一格中的颜色，翻到下一页，填涂上去。

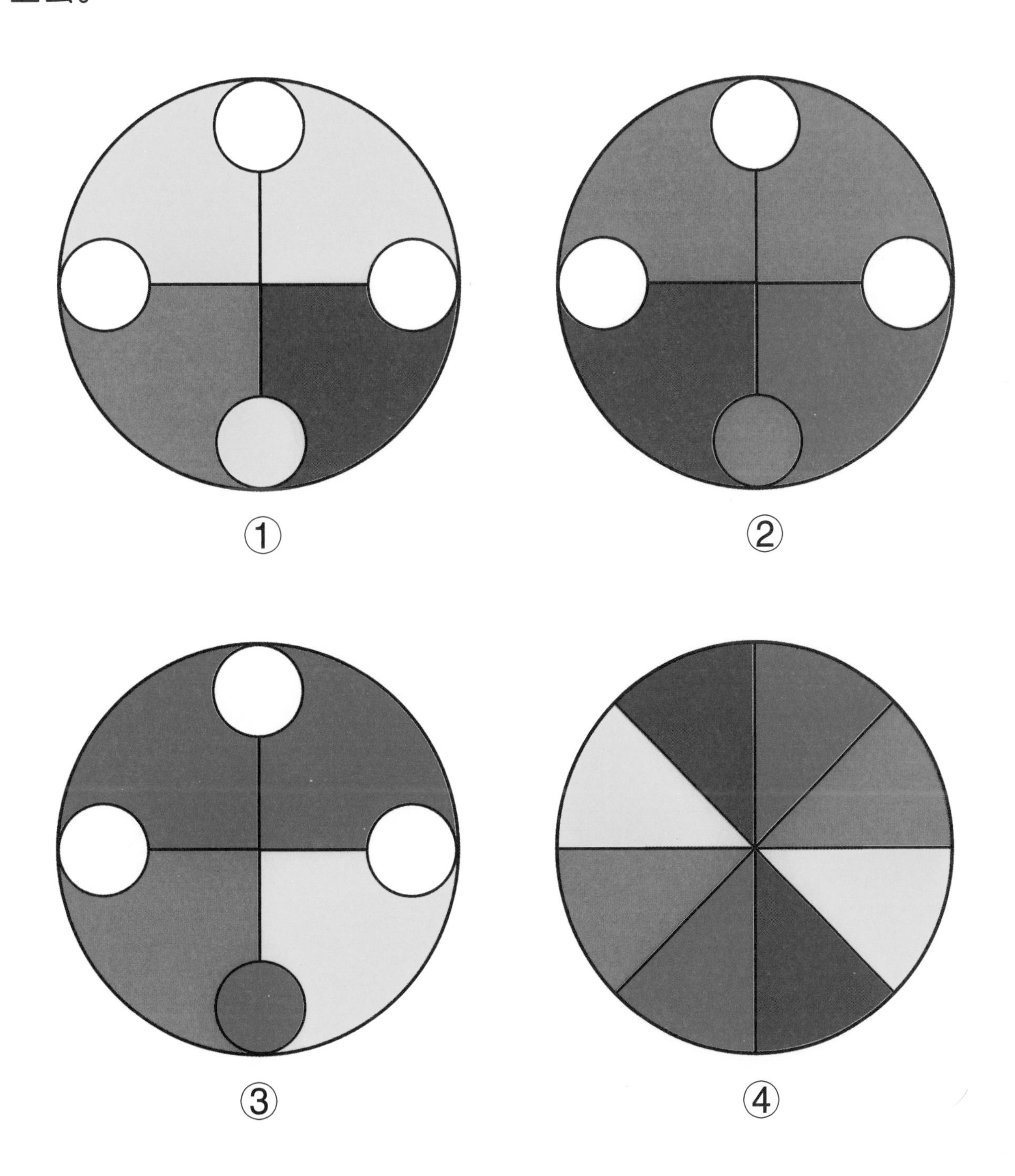

游戏难度 ★★★★☆

给下列曼陀罗图案涂上颜色。

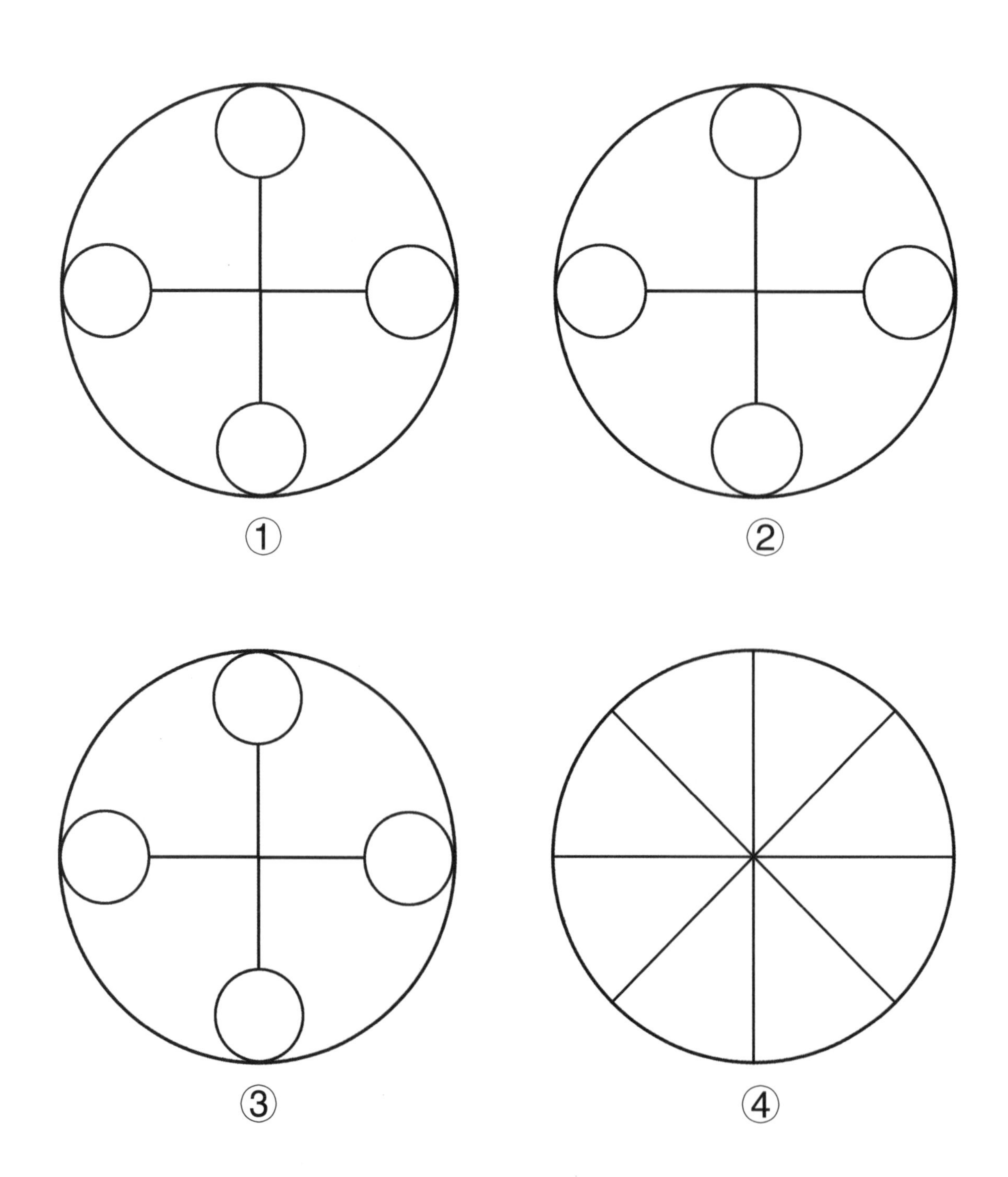

曼陀罗涂色（四）

看下面的曼陀罗图案，记住每一格中的颜色，翻到下一页，填涂上去。

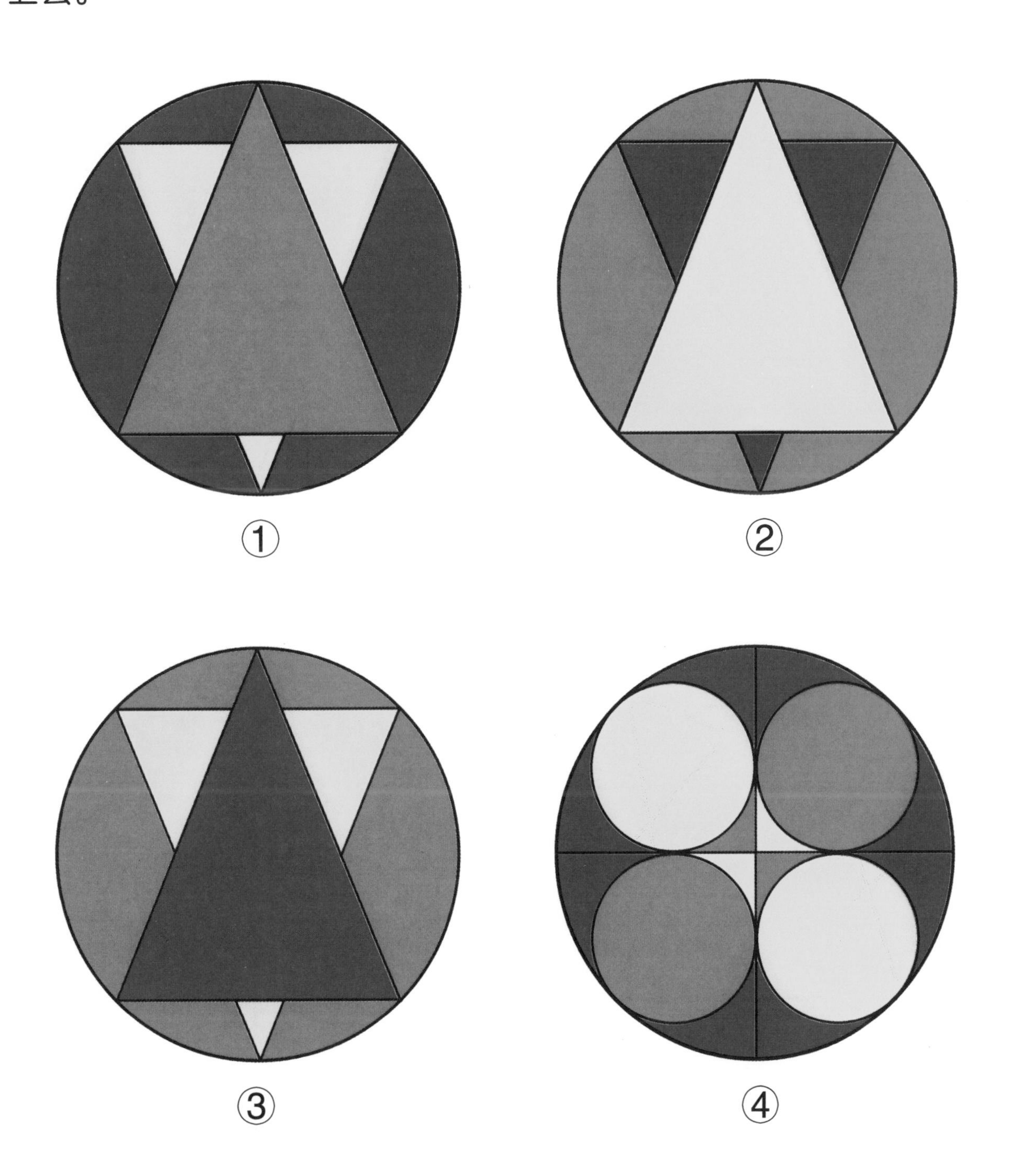

游戏难度 ★★★☆☆

给下列曼陀罗图案涂上颜色。

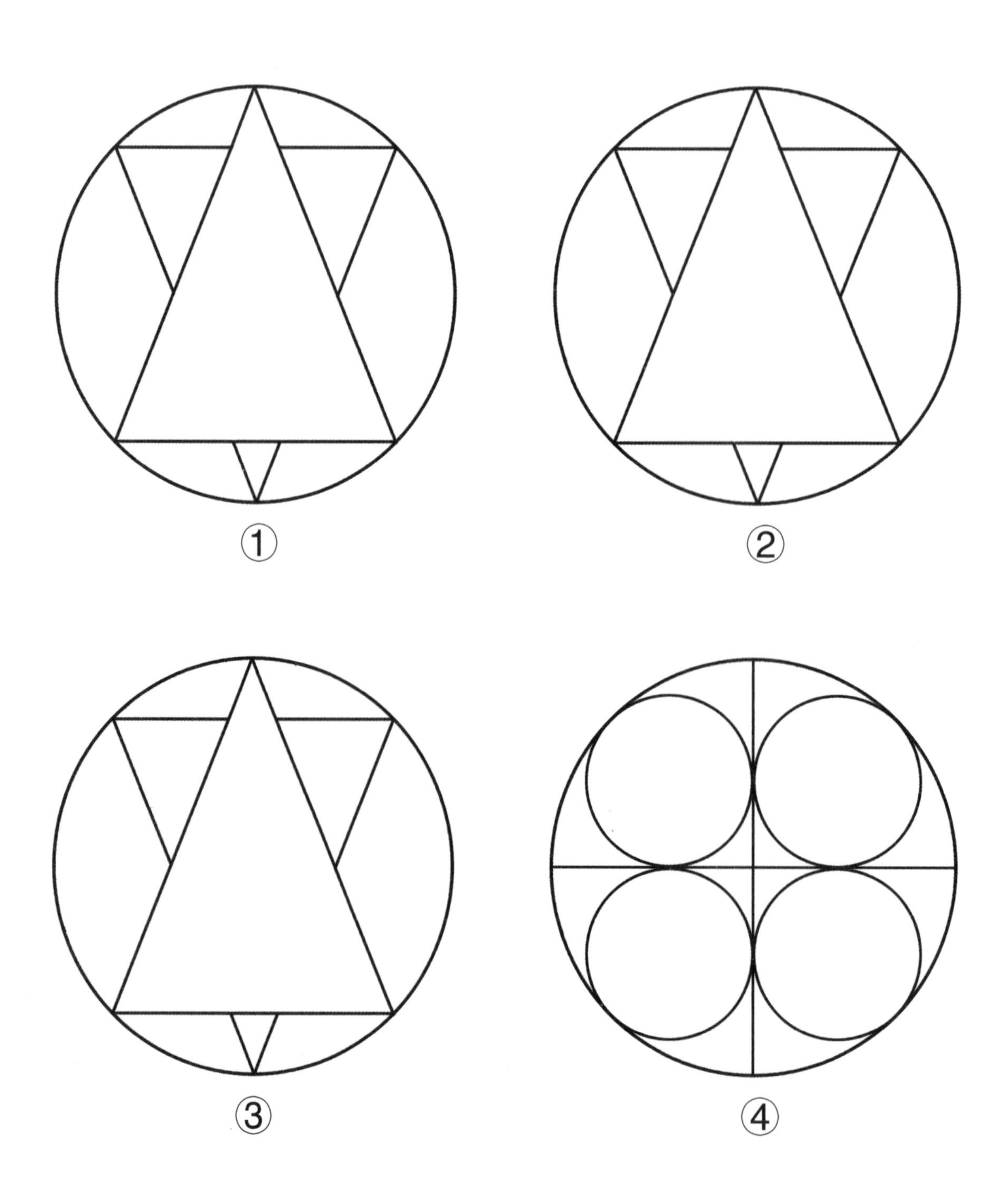

曼陀罗涂色

曼陀罗涂色（五）

看下面的曼陀罗图案，记住每一格中的颜色，翻到下一页，填涂上去。

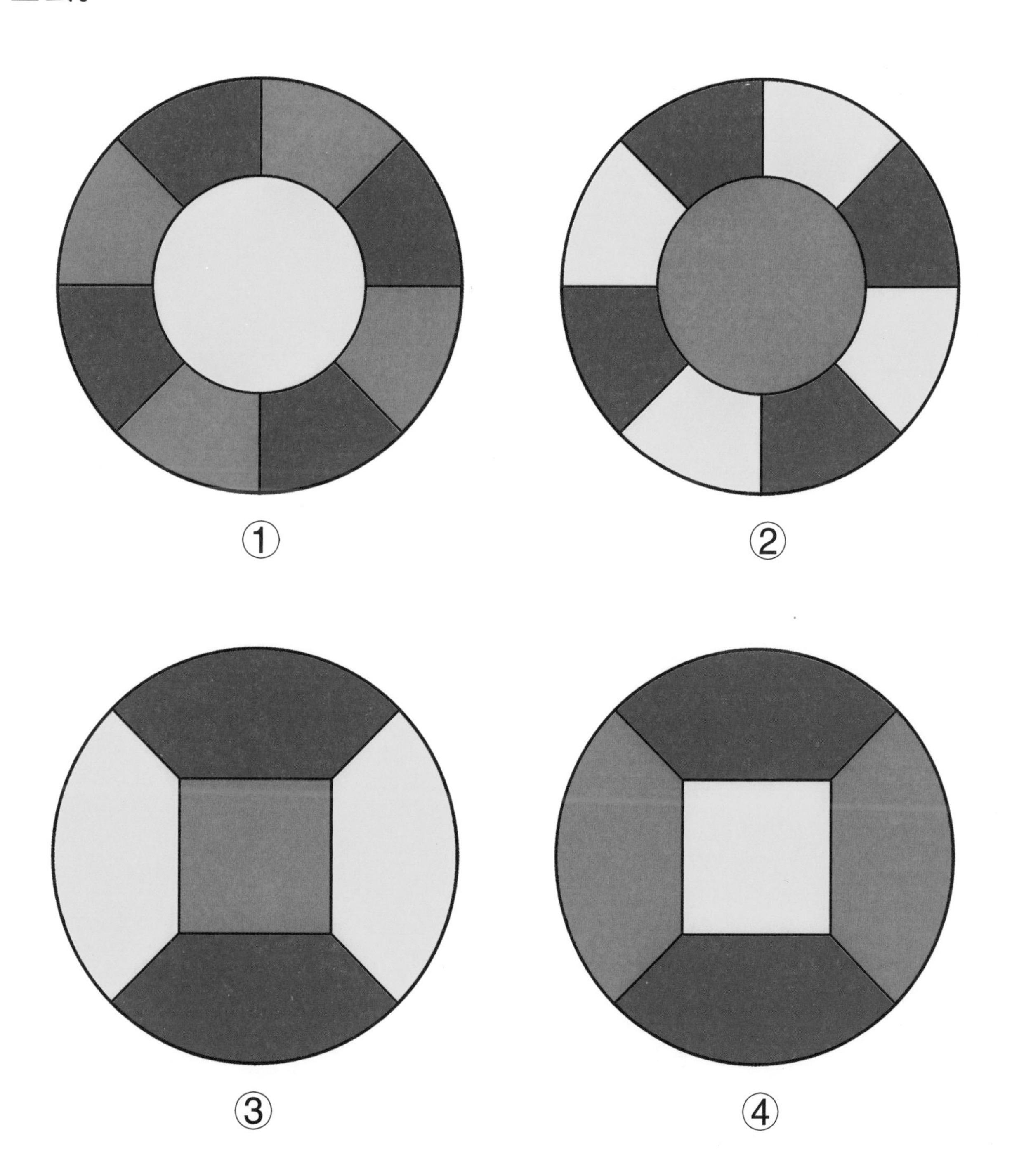

游戏难度 ★★★☆☆

给下列曼陀罗图案涂上颜色。

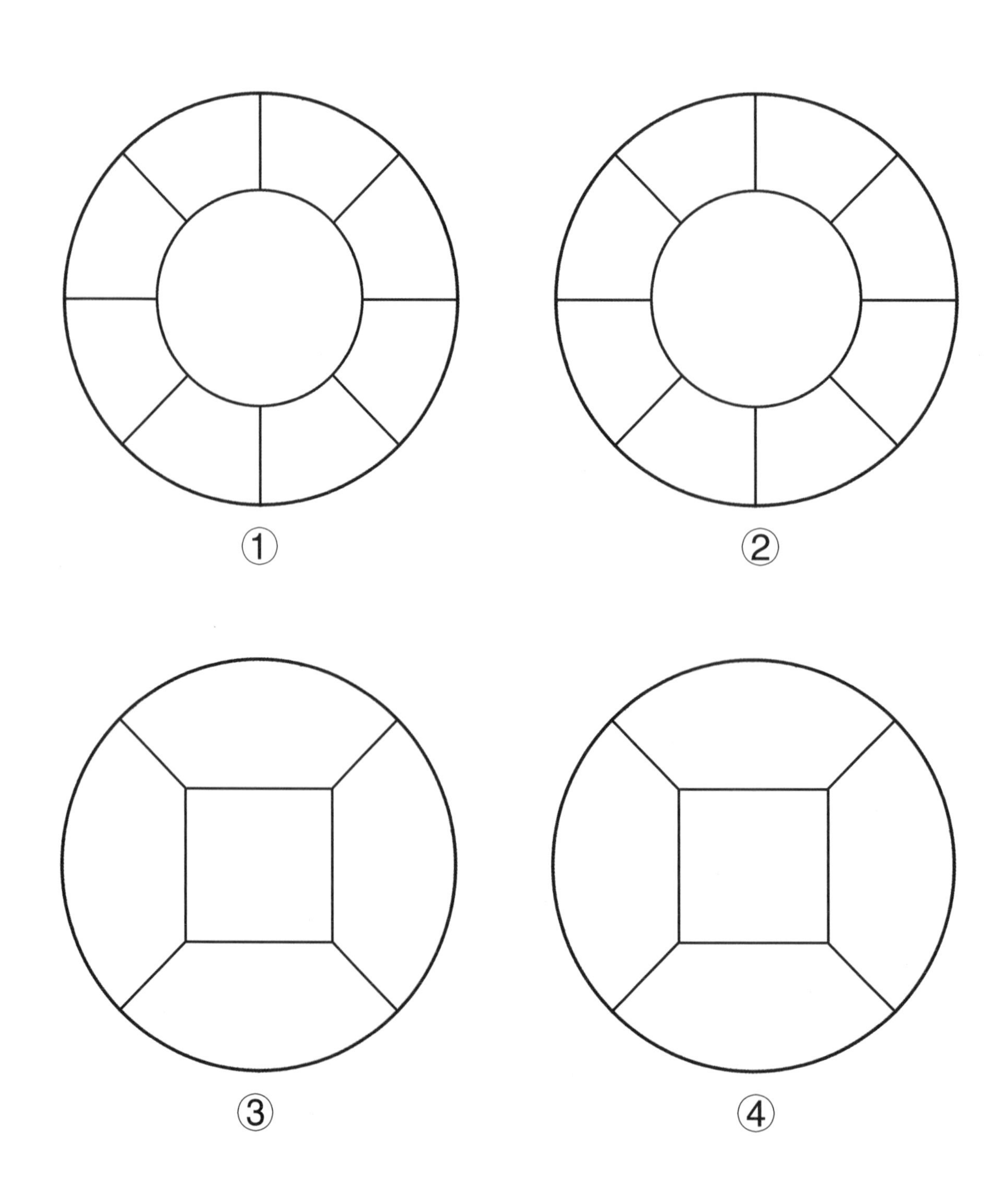

曼陀罗涂色（六）

看下面的曼陀罗图案，记住每一格中的颜色，翻到下一页，填涂上去。

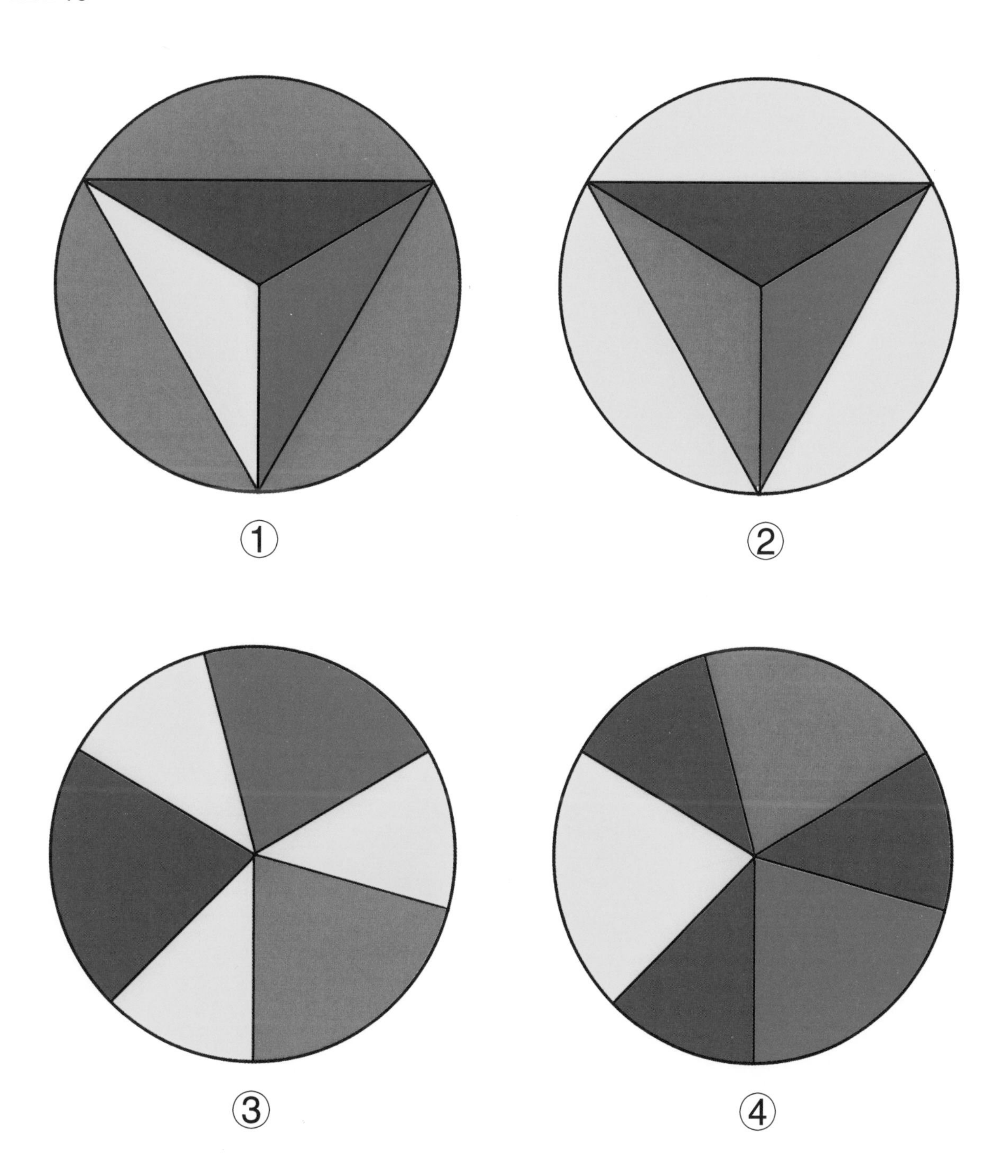

游戏难度 ★★★★☆

给下列曼陀罗图案涂上颜色。

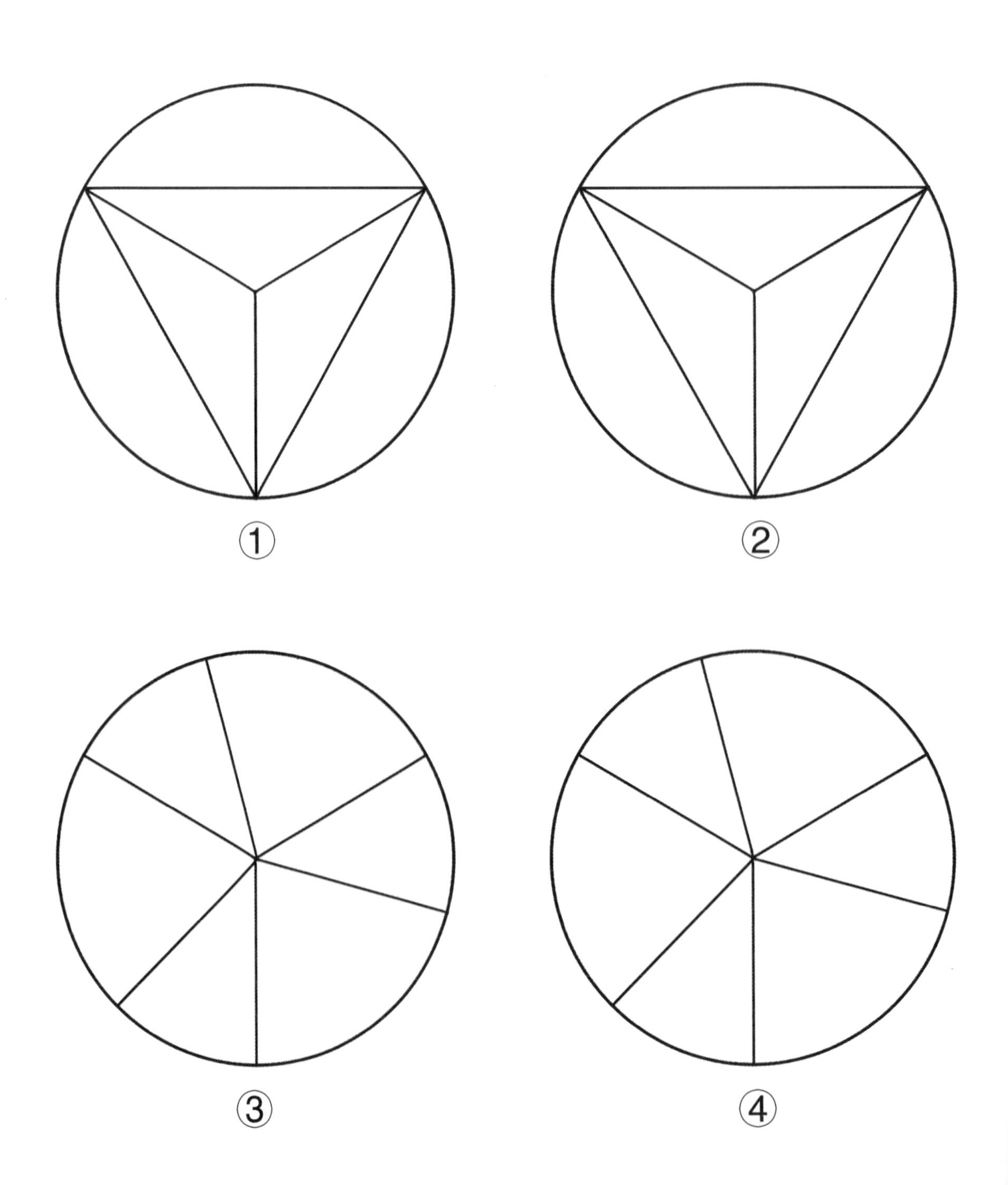

曼陀罗涂色（七）

看下面的曼陀罗图案，记住每一格中的颜色，翻到下一页，填涂上去。

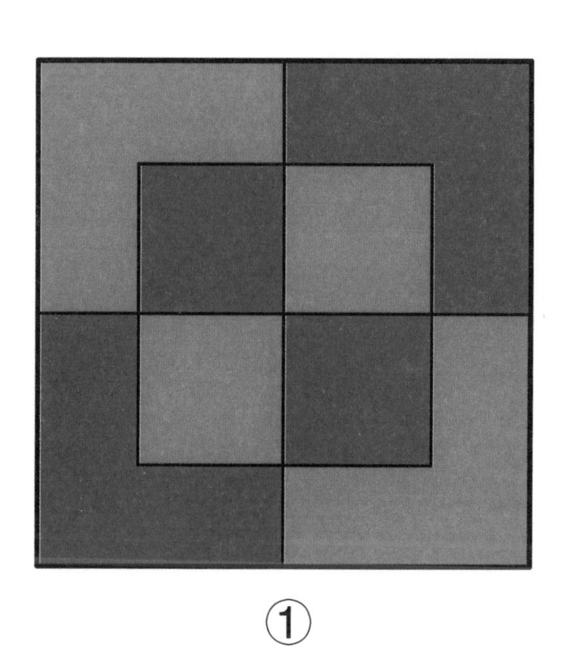

①

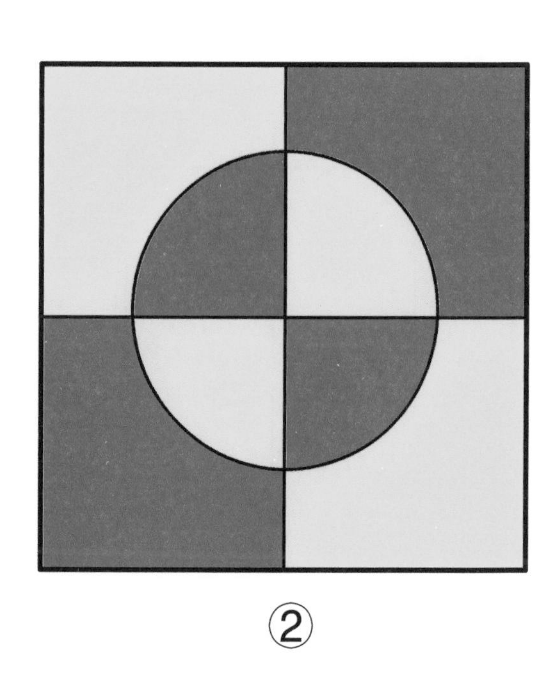

②

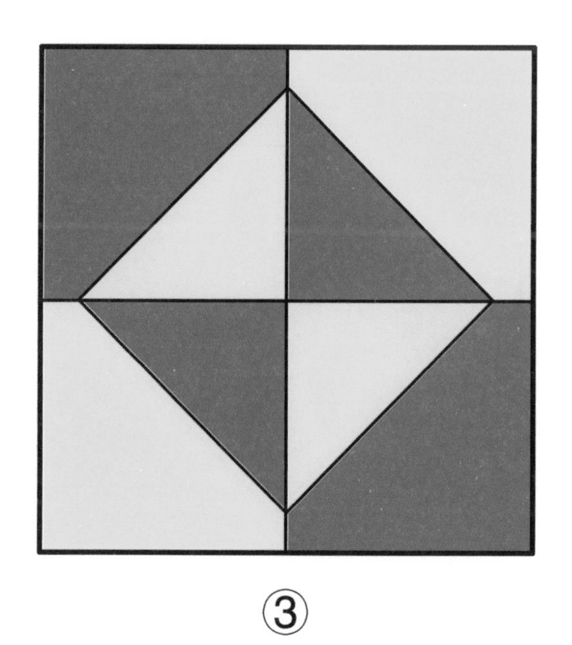

③

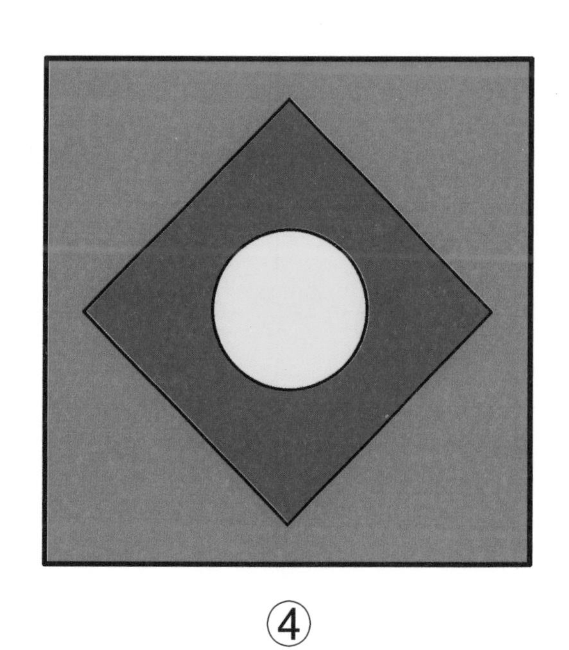

④

游戏难度 ★★★☆☆

给下列曼陀罗图案涂上颜色。

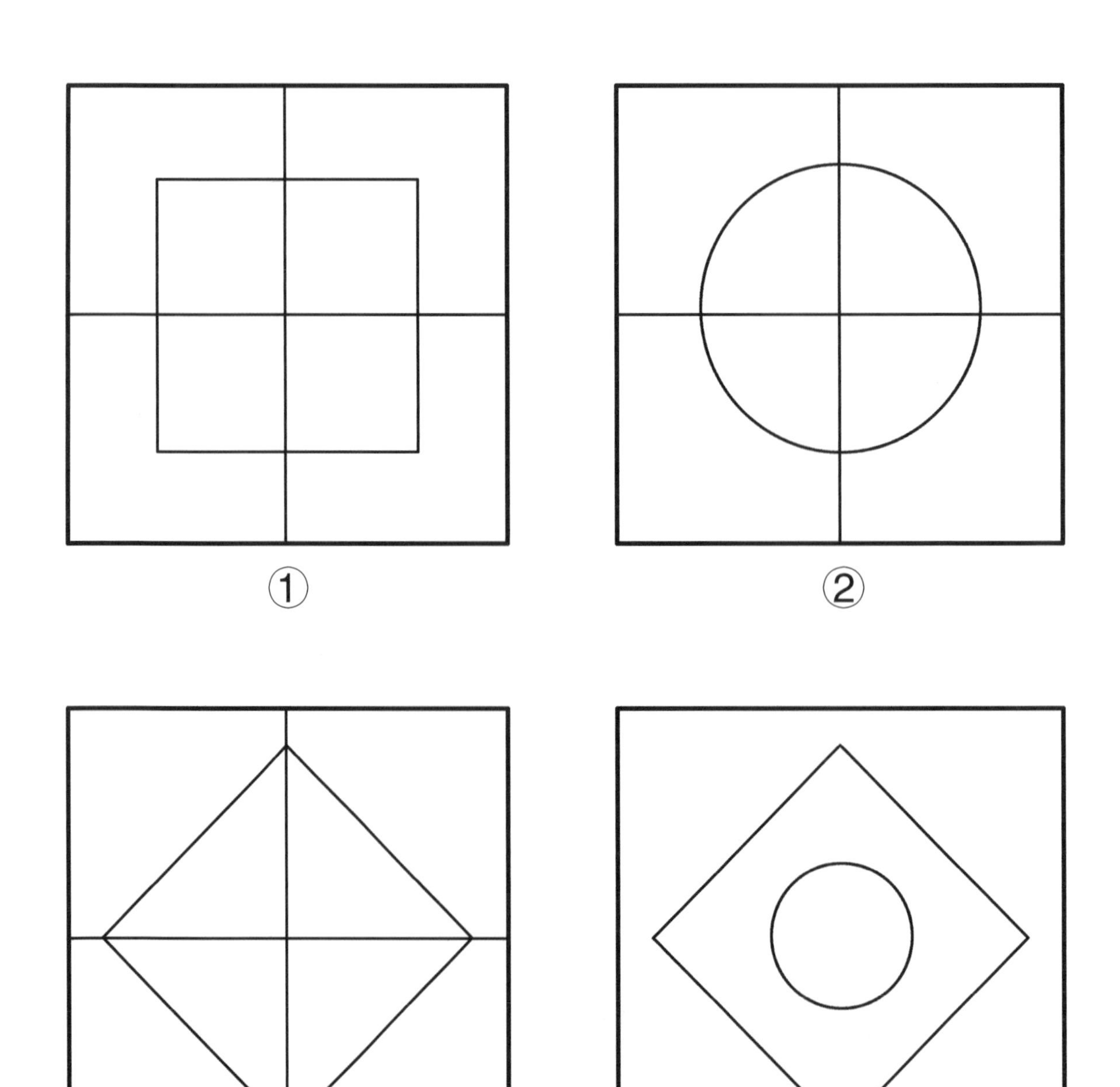

曼陀罗涂色（八）

看下面的曼陀罗图案，记住每一格中的颜色，翻到下一页，填涂上去。

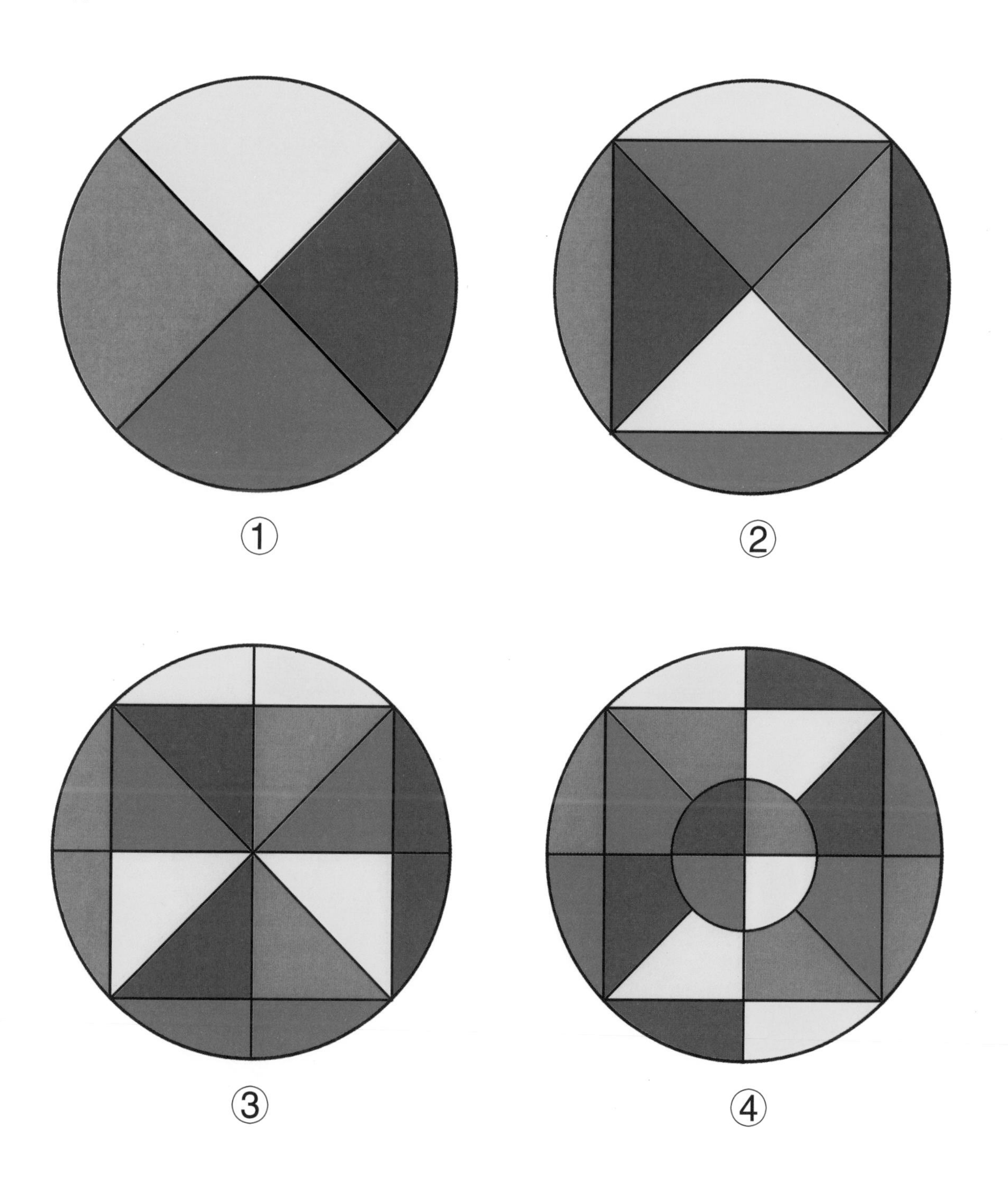

游戏难度 ★★★★★

给下列曼陀罗图案涂上颜色。

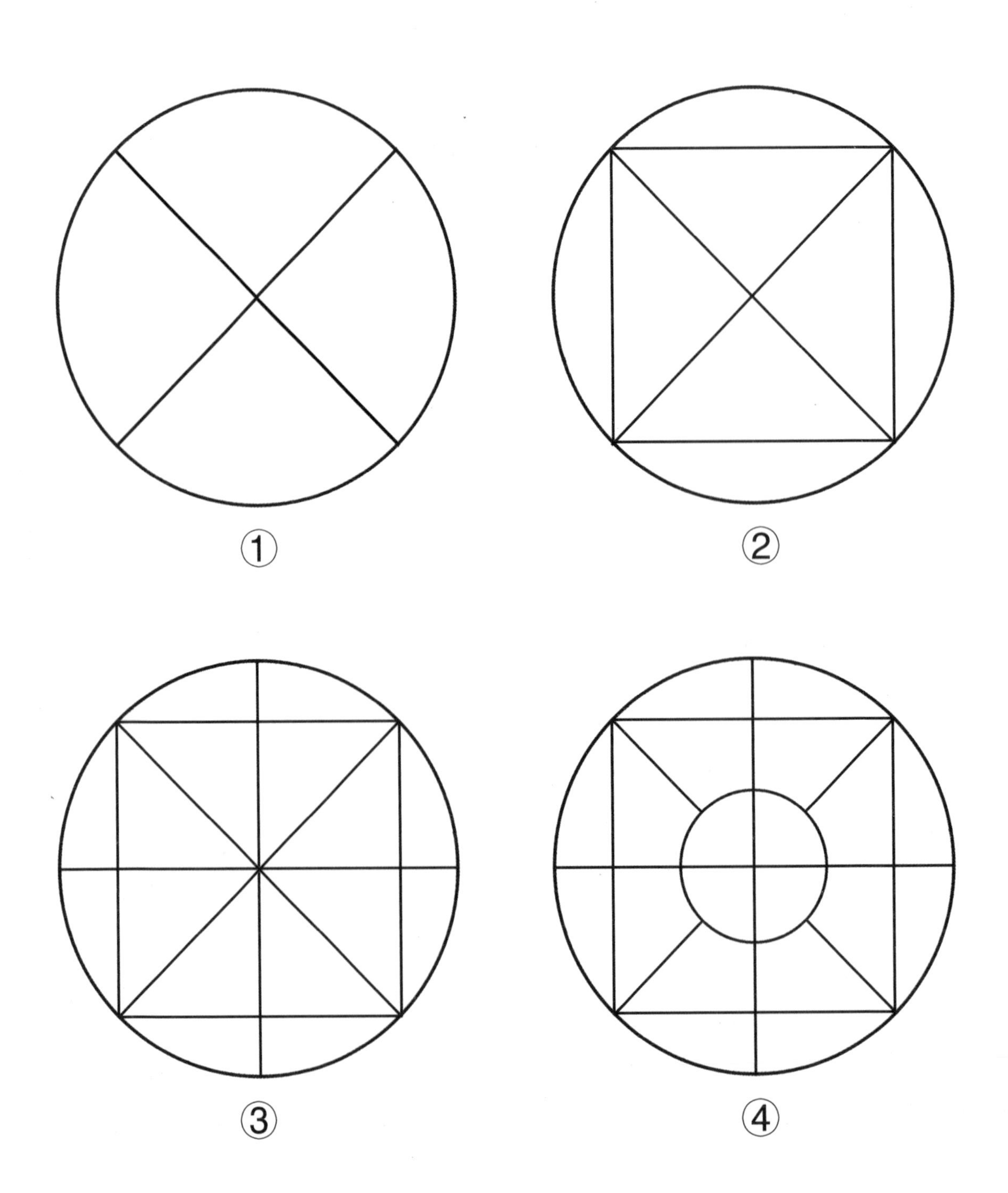